Nicotine and Other Tobacco Compounds in Neurodegenerative and Psychiatric Diseases

Nicotine and Other Tobacco Compounds in Neurodegenerative and Psychiatric Diseases

Overview of Epidemiological Data on Smoking and Preclinical and Clinical Data on Nicotine

Emilija Veljkovic

Wenhao Xia

Blaine Phillips

Ee Tsin Wong

Jenny Ho

Alberto Oviedo

Julia Hoeng

Manuel Peitsch

Academic Press is an imprint of Elsevier
125 London Wall, London EC2Y 5AS, United Kingdom
525 B Street, Suite 1800, San Diego, CA 92101-4495, United States
50 Hampshire Street, 5th Floor, Cambridge, MA 02139, United States
The Boulevard, Langford Lane, Kidlington, Oxford OX5 1GB, United Kingdom

Library of Congress Cataloging-in-Publication Data
A catalog record for this book is available from the Library of Congress

British Library Cataloguing-in-Publication Data
A catalogue record for this book is available from the British Library

ISBN 978-0-12-812922-7

For information on all Academic Press publications
visit our website at https://www.elsevier.com/books-and-journals

Publisher: Nikki Levy
Acquisition Editor: Natalie Farra
Editorial Project Manager: Tracy Tufaga
Senior Production Project Manager: Priya Kumaraguruparan
Cover Designer: Miles Hitchen

Typeset by SPi Global, India

Contents

Disclaimer

Chemical structures in 2D were generated using MarvinView 14.9.8.0 from where they were exported in .png format. The views expressed herein are those of the individual authors/researchers and do not necessarily reflect those of Philip Morris International Inc. and/or its subsidiaries.

Acknowledgments

Authors would like to thank Pavel Pospíšil, PhD for preparing the molecule structures used in this book. The authors would like to thank Stephanie Boue, PhD for preparing the illustrations used in this book.

Foreword

Basic and clinical research on therapeutics targeting the nicotinic acetylcholine receptors (nAChRs) was greatly accelerated with the discovery of epibatidine from the skin of the Ecuadorian poison arrow frog *Epipedobates tricolor* by John Daly in 1992.[1] The compound, found to be an analgesic more powerful than morphine in early pharmacological profiling, was a potent ligand at virtually all the various subtypes of nAChRs (α4β2, α4β2α5, α4α6β2β3, α6β2β3, α6β2, α3β4, α3β4β3, α4β3α5, and α7).[2] On the heels on Daly's discovery, biotech and pharmaceutical companies began decades of nicotinic drug discovery and development programs targeting primarily brain diseases. While much of the early pharmaceutical science focused on the identification of novel pain therapeutics, the first pharmaceutical drug designed to target nAChRs other than nicotine was Chantix/Champix (varenicline), which was developed by Pfizer and approved as an aid to smoking cessation by both the Food and Drug Administration (FDA) and the European Medicines Agency (EMA) in 2006. The focus on brain diseases was particularly driven by epidemiological studies suggesting that smokers who suffer from neuropsychiatric conditions like schizophrenia, Tourette's syndrome, attention deficit hyperactivity disorder (ADHD), and even Alzheimer's disease (AD) may experience some level of alleviation of their disease symptoms from the activity of smoking. Furthermore, some conditions or diseases like Parkinson's disease (PD), ulcerative colitis (UC), preeclampsia, and symptomatic gallstones have been labeled as "diseases of nonsmokers" because of the markedly lower incidences of these diseases in the smoking population compared with nonsmokers. Given that the best-studied constituent of tobacco smoke is nicotine, much of the abovementioned effects of smoking have been attributed to nicotine's effects on its receptors.

Given the context of its known benefits, it might then be reasonable to consider nicotine in a therapeutic context. However, while nicotine delivered to the body other than by smoking (transdermal patch, gum, etc.) to aid smoking cessation is not considered to be dangerous, nicotine itself has not historically been believed to be useful as a drug, due to its short plasma half-life and its potential for adverse effects (it is a stimulant with attendant addiction potential, it has vasoconstrictor properties with adverse cardiovascular effects, and it can accelerate tumor growth). Because nicotine and several other naturally occurring alkaloids (epibatidine, cytisine, etc.) interact at nicotinic receptors nonselectively, drug discovery efforts over the past 25 years have focused on the development of novel subtype-selective drugs targeting nAChRs in order to more precisely leverage the neuronal nicotinic receptors in disease states.

As such, the following book by Veljkovic et al. entitled *Nicotine and other tobacco compounds in neurodegenerative and psychiatric diseases: Overview of epidemiological data on smoking and preclinical and clinical data on nicotine* elegantly details the role of nicotinic receptors in diseases of the brain and highlights other tobacco constituents, decoupled from combusted tobacco products, which can also contribute effects observed among smokers with these conditions. First, the authors undertake a systematic approach of the evidence of nicotine's involvement in multiple brain disorders and diseases, evaluating the symptoms and epidemiology, the molecular mechanisms involved, and the impact of smoking and nicotine in those diseases. After that, a review of the mechanisms of nicotine's biological effects and of other constituents of tobacco potentially involved in neuroprotection is presented.

Such an involved and updated treatment and analysis of nicotine's effects in diseases of the brain is useful because the past 15 years have witnessed multiple products targeting nicotinic receptors fail during clinical evaluation, either at proof-of-efficacy phase 2 trials or in large phase 3 trials. This includes several attempts by pharmaceutical companies (big and small) to develop nicotinic products for pain, AD, PD, schizophrenia, cognitive deficits in schizophrenia, ADHD, and major depressive disorders. There are multiple reasons that have been attributed to the difficulty of translating solid preclinical and epidemiological evidence for drugs acting at nicotinic receptors to sustained clinical effects in the central nervous system (CNS). The nAChR class of receptors is complex and often displays a disconnect between pharmacokinetic and pharmacodynamics effects, which has made the identification of doses that should be optimal for a given disease indication very challenging. CNS drug development is difficult to begin with, and success rates in therapeutics in CNS are known to be poor relative to other areas like cardiovascular, hematology, and other broad classifications of disease. And even after the above considerations, the design of sufficiently subtype-selective ligands that act at neuronal nicotinic receptors remains problematic.

Despite this difficult history, the benefits to addressing the above challenges of nicotinic drug development remain high for the patient, for the companies that bring forth those drugs, and for the underlying science, which depends on the growing understanding and feedback from human experience of modulating these receptors to inform new areas of clinical potential for nAChRs. While mechanism-based tolerability has occasionally presented a dose-limiting challenge in drug development of drugs targeting certain nAChR subtypes, life-threatening adverse events have rarely been the cause of clinical failure of experimental nicotinic drugs. Therapeutics targeting nicotinic receptors that are sufficiently well tolerated and selective (both within the proteome and within the gene family of nicotinic receptors) will always be compelling to evaluate for clinical utility in areas of high unmet need like Parkinson's disease, schizophrenia, and other central or peripheral nervous system indications. Similarly, alternative delivery forms of nicotine (or other naturally occurring nicotinic modulators) that minimize its side effects may also find utility in certain indications. Several new developments have occurred and are occurring that pave the path to renewed interest in nicotinic drug development.

First, receptor subtypes are and have been emerging that are strongly implicated in animal proof-of-concept studies in indications both previously associated with nAChRs and ones newly associated with this gene family. Nicotinic agonists acting at α6β2-containing receptors are of interest to modulate the motor effects of PD, as it is known that this receptor subtype degenerates with progression of disease. These same receptors have been studied in elegant gain-of-function and loss-of-function mutants to confirm the analgesic role of the α6 subunit in mouse models of neuropathic and inflammatory pain. Agonists targeting α5-containing receptors (α4β2α5 and α3β4α5) are being developed for nicotine addiction, as α5 nAChRs have been shown to mediate aversive responses to nicotine consumption, while antagonists of α5-containing receptors may be useful for the treatment of small-cell and nonsmall-cell lung cancer. And α9α10 selective agonists are considered to have therapeutic protective value in conditions such as noise-induced hearing loss, tinnitus, and auditory processing disorders; α9α10 selective antagonists, on the other hand, have been proposed for the treatment of chronic pain. For many of these emerging receptors, selective tool peptides are known. In these cases, either the peptides need to be made more druggable, or small-molecule substitutes of the peptides need to emerge to advance therapeutics targeting these receptors. Despite these opportunities, more drug discovery innovation is needed to enhance the probability of success in progressing ligands that act at the nAChR class of receptors.

Second, there is greater awareness that more selective ligands are needed for the receptors most known to positively impact the neurodegenerative and psychiatric diseases and covered in this book. In the case of nicotinic drugs, dose-limiting adverse events (typically on-target, mechanism-related tolerability issues) have been a problem, but sometimes, separation from other proteomic targets, notably 5-hydroxytryptamine (5-HT_3), can be problematic as well. 5-HT_3 belongs to the same Cys-loop superfamily of ligand-gated ion channels (LGICs) as nAChRs, and they are closely related by homology. Forum Pharmaceuticals (formerly known as En Vivo Pharmaceuticals) suffered devastating clinical setbacks in its industry-leading α7 agonist program for AD and cognitive impairment in schizophrenia this past year when encenicline (EVP-6124) failed in both phase 3 trials. While replicating the efficacy of its successful phase 2 programs in these CNS diseases was the major problem for Forum, it is clear that the relatively equal potency of encenicline at both α7 and 5-HT_3

receptors did complicate the clinical programs, particularly in AD, where the consequential rare but severe gastrointestinal side effects in elderly AD patients caused an FDA-mandated clinical hold on the program.

Third, greater translational understanding of the underlying science of some of these psychiatric conditions is needed. Two papers recently published underscore this point. First, Yin et al.[3] assert that mice deficient in the α7 nAChR manifest no consistent neuropsychiatric and behavioral phenotypes. Consistent with this perspective, a translational meta-analysis of rodent and human studies of α7 nicotinic agonists for cognitive deficits in neuropsychiatric disorders determined that targeting the α7 nAChR with agonists may not be a robust treatment for cognitive dysfunction in schizophrenia or AD, suggesting a more clear understanding is needed of the translational gap for therapeutics targeting the α7 nAChR.[4]

Finally, for the nicotinic receptor subtypes that have been most well studied, novel clinical indications need to be established. Neuroinflammation, for instance, has become a hot area of research over the past 20 years, necessitating a journal dedicated to the publication of research exclusive to this topic (Journal of Neuroinflammation, established in 2004). The start of the ramp-up in neuroinflammation research and publications coincides with the publication of seminal papers by Kevin Tracey published in the early 2000s describing the role of the endogenous cholinergic antiinflammatory pathway, involving the α7 nicotinic receptor, in the mitigation of an overactive inflammatory response.[5] This approach has found particular utility in neuroinflammation.[6] However, despite two decades of increased preclinical understanding of this mechanism and multiple examples of preclinical proof-of-concept animal studies, not a single pharmacotherapeutic test of this antiinflammatory hypothesis has been executed in a clinical setting. In recognition of this opportunity, a small but growing number of companies are now involved in commercialization of this pathway in clinical indications like cough, arthritis, and neuroinflammatory indications of traumatic brain injury and postoperative cognitive dysfunction (POCD). In the case of POCD, an opportunity exists to intervene prophylactically with a drug in advance of a known trigger of inflammation (surgery) at a discreet point in time, which presents a novel modality to *prevent* inflammatory cognitive impairment and to treat the inflammation once it has initiated.

While past efforts in the clinical development of nicotinic drugs have proved challenging, the application of more learning about the basic underlying disease pathology and optimal dosing of nicotinic drugs, combined with the inherent druggability of this class of receptors, should provide more clinical success stories. Thus, the outlook for drug development based on therapeutics targeting nAChRs remains positive, with the coming decade set to provide multiple examples of value realization in this field. And as we keep an open mind about the future of drug development at nicotinic receptors, we must anticipate that nicotine delivered from vaping or e-cigarettes may over time prove to discern whether the effects of smoking observed in neurodegenerative diseases such as PD are accrued to nicotine versus other constituents in tobacco. The following book, authored by those intimately familiar with diseases that involve nAChRs and with research based in the pharmacology and drug delivery of nicotine and nicotinic ligands, stands as valuable reading to contextualize advances in this field.

Daniel Yohannes
Yohannes Pharmaceutical Consulting

Introduction

Smoking cigarettes is causally linked to a number of serious diseases. These include cancers of the larynx, lung, bladder, and digestive tract; chronic obstructive pulmonary disease (COPD); and a number of cardiovascular diseases, such as aortic aneurysm, stroke, and ischemic heart disease.[7] In addition, the 2014 U.S. Surgeon General Report states that smoking is causally associated with inflammation and impaired immune function, and that smokers are at higher risk of developing pneumonia, tuberculosis, and other airway infections.[8] Clearly, the best way to avoid harm from smoking is to never start, and for smokers to quit.

In 1976, Michael Russell wrote that *people smoke for the nicotine but they die from the tar.*[9] Indeed, most of the harm caused by cigarette smoking is due to the continued inhalation of toxicants, the majority of which are generated by the combustion of tobacco. It is widely recognized that *nicotine itself is not a highly hazardous drug*[10] and that *most of the harm caused by smoking arises not from nicotine but from other components of tobacco smoke.*[11,12]

Whereas the risks of serious diseases is clearly increased by smoking, the relationship between smoking (and smokeless tobacco consumption) and neurodegenerative and psychiatric diseases is more differentiated and complex. For instance, epidemiological evidence shows a lower incidence of Parkinson's disease (PD) in smokers.[13] While cigarette smoking is recognized as a risk factor for the development of multiple sclerosis (MS),[14] the use of Swedish snus[a] does not appear to be associated with increased MS risk.[15] Despite mixed epidemiological data, smoking is recognized as a risk factor for Alzheimer's disease (AD).[16,17] However, there is clinical evidence suggesting that acute nicotine administration improves recall, visual attention, and mood in AD patients.[18,19] There is evidence that smoking may cause both attention deficit hyperactivity disorder (ADHD) and depression to some extent, and equally, it has been reported that subjects with ADHD or depression are more likely to smoke.[20,21] Finally, patients describe partial attenuation of symptoms, such as cognitive improvement, anxiety reduction, and stress relief with smoking (e.g., in schizophrenia, anxiety, and depression).

---

a. A smokeless moist powder tobacco product in pouch form typically placed under the upper lip for extended periods, delivering a dose of nicotine equivalent to cigarette smoking, but without the toxic substances generated by cigarette combustion.

Nicotine is the major alkaloid in tobacco and exerts its pharmacological effect by binding to various forms of nicotinic acetylcholine receptors (nAChRs) located both in the central nervous system (CNS) and at the periphery. The neuroprotective effects of nicotine have been evaluated by others in *in vitro* and *in vivo* studies and in clinical trials using various routes of administration. A number of scientific reports also suggest that nicotine exerts antiinflammatory and antiapoptotic effects.[22–24] Furthermore, the effects of nicotine on various neurotransmitter systems may provide a possible explanation for the partial relief of symptoms reported by patients suffering from some of the neurodegenerative diseases mentioned above.[25] However, the safety of nicotine use must be evaluated carefully when considering its therapeutic properties. Its addictive potential, acute toxicity and impact on brain development are just some of the undesirable effects in this context.

The addictive potential of smoking is attributed primarily to nicotine.[26] However, when decoupled from smoking tobacco products, the addictiveness of nicotine may be lower.[27,28] Even though nicotine-containing products other than cigarettes are still addictive, decades of research and use have shown that nicotine replacement therapy (NRT) products do not appear to have significant potential for abuse or dependence.[29,30]

At high doses, at least 500 mg, nicotine can induce acute toxicity when ingested.[31] At the lower doses typically seen in cigarettes and NRT products, nicotine can cause temporary discomfort (e.g., nausea and vomiting) in nontolerant individuals, but is not associated with clinically significant toxicity.[32]

While prolonged use of nicotine at doses delivered by existing nicotine and tobacco products may not be of particular toxicological concern in adults, it should be avoided during pregnancy as nicotine appears to cause several adverse effects during pregnancy. Available data are mainly from epidemiological studies on Swedish snus consumed by pregnant women. Swedish snus delivers nicotine in amounts similar to cigarettes but in the absence of combustion-related smoke toxicants and with reduced levels of tobacco-specific nitrosamines when compared with conventional smokeless tobacco products.[33] It can therefore be assumed that adverse effects observed in pregnant women consuming Swedish snus can be primarily attributable to nicotine. The mean birth weight of newborns of snus-consuming women was reduced compared with nontobacco users but to a lesser extent than compared with smokers.[34] Preterm deliveries were increased in snus users and smokers, while preeclampsia was reduced in smokers but increased in snus users.[35] Use of snus during pregnancy was associated with a higher risk of stillbirth.[36] Furthermore, there is growing concern and evidence that nicotine exposure during pregnancy alters the normal brain development of the child. Indeed, *in utero* and postnatal exposure to nicotine has been reported as a risk factor for childhood and adolescent cognitive and neurobehavioral challenges, such as ADHD.[37,38]

The results of clinical trials conducted with nicotine in neurodegenerative diseases are often difficult to interpret and reconcile with the epidemiological data linking smoking with the incidence of these diseases. While the results of some trials are encouraging, with patients reporting significant improvements, others are less so. This may be due to two reasons. First, delivering nicotine via repeated inhalation throughout the day (i.e., smoking), 24 h patches, or gum results in different pharmacokinetic profiles that may drive different responses. Second, nicotine is most probably not the only compound in cigarette smoke responsible for these effects. Indeed, tobacco alkaloids with chemical structures similar to nicotine, such as anatabine and anabasine, may act together with nicotine to exert pharmacological effects. Furthermore, cembranoids, nicotine metabolites, monoamine oxidase (MAO) inhibitors, and other tobacco constituents may contribute to these effects.

The primary purpose of this book is to provide a broad overview of the epidemiological evidence linking smoking with several neurodegenerative and psychiatric diseases and to complement this with a summary of the results of both clinical and preclinical studies conducted with nicotine in these diseases. The second part of this book describes how nAChRs, nicotine, and several less well-studied tobacco-derived compounds may contribute to neuronal protection.

Search Methods

To investigate the epidemiological association between cigarette smoking and various neurodegenerative and psychiatric diseases, we searched PubMed using the following keywords:

- PD: "Cigarette smoking" AND "Parkinson's Disease" AND "Epidemiology" with filter "Species = Human"
- AD: "Cigarette smoking" AND "Alzheimer's Disease" AND "Epidemiology" with filter "Species = Human"
- MS: "Cigarette smoking" AND "Multiple sclerosis" AND "Epidemiology" with filter "Species = Human"
- Schizophrenia: "Cigarette smoking" AND "Schizophrenia" AND "Epidemiology" with filter "Species = Human"
- Tourette's Syndrome: "Cigarette smoking" AND "Tourette's Syndrome" AND "Epidemiology" with filter "Species = Human"
- ADHD: "Cigarette smoking" AND "Attention deficit hyperactivity disorder" AND "Epidemiology" with filter "Species = Human"
- Depression: "Cigarette smoking" AND "Depression" or "Depressive disorder" AND "Epidemiology" with filter "Species = Human"
- Anxiety: "Cigarette smoking" AND "Anxiety" AND "Epidemiology" with filter "Species = Human"

The search results included articles published through Mar. 2017. Articles that were not written in English or were irrelevant to the topic were excluded. Of the remaining results, we first reviewed publications describing a meta-analysis of the historical studies. Where additional relevant studies were identified that were subsequent to the latest meta-analyses reported, they were also included. We then consolidated the data into disease-specific tables. Since we performed the searches only in PubMed, studies published in articles not referenced in PubMed are not represented.

To collect information on past and ongoing clinical trials using nicotine as an intervention in neurodegenerative and neuropsychiatric diseases, we performed searches in both PubMed and Clinicaltrials.gov, using the following keywords:

PubMed search:

- PD: "Nicotine" AND "Parkinson's Disease" with filters "Species = Human" and "Article type = Clinical trial"

- AD: "Nicotine" AND "Alzheimer's Disease" with filters "Species = Human" and "Article type = Clinical trial"
- MS: "Nicotine" AND "Multiple sclerosis" with filters "Species = Human" and "Article type = Clinical trial"
- Schizophrenia: "Nicotine" AND "Schizophrenia" with filters "Species = Human" and "Article type = Clinical trial"
- Tourette's Syndrome: "Nicotine" AND "Tourette's Syndrome" with filters "Species = Human" and "Article type = Clinical trial"
- ADHD: "Nicotine" AND "Attention deficit hyperactivity disorder" with filters "Species = Human" and "Article type = Clinical trial"
- Depression: "Nicotine" AND "Depression" or "Depressive disorder" with filters "Species = Human" and "Article type = Clinical trial"
- Anxiety: "Nicotine" AND "Anxiety" with filters "Species = Human" and "Article type = Clinical trial"

Clinicaltrials.gov search:

We searched for studies with "nicotine" as a keyword. The results were then sorted "by topic," and entries under each of the pertinent diseases were reviewed.

We extracted the relevant results from both database searches, up through Mar. 2017. The data were used to build disease-specific tables.

In order to retrieve the information on other nicotinic receptor agonists that entered clinical trials or are on the market, we first searched the International Union of Basic and Clinical Pharmacology (IUPHAR) database and "AdisInsight" database to identify the nicotinic receptor agonists. The name of the identified compounds was then used to search against "Clinicaltrials.gov" and "PubMed" to review their clinical development status. Compounds that did not appear in these databases were not included in the book.

To review past studies demonstrating nicotine's effects in animal models for various neurodegenerative/neuropsychiatric diseases, we searched PubMed using the following keywords:

- PD: "Nicotine" AND "Parkinson's Disease" with filter "Species = Other animals"
- AD: "Nicotine" AND "Alzheimer's Disease" with filter "Species = Other animals"
- MS: "Nicotine" AND "Multiple sclerosis" with filter "Species = Other animals"
- Schizophrenia: "Nicotine" AND "Schizophrenia" with filter "Species = Other animals"
- Tourette's Syndrome: "Nicotine" AND "Tourette's Syndrome" with filter "Species = Other animals"
- ADHD: "Nicotine" AND "Attention deficit hyperactivity disorder" with filter "Species = Other animals"

- Depression: "Nicotine" AND "Depression" or "Depressive disorder" with filter "Species = Other animals"
- Anxiety: "Nicotine" AND "Anxiety" with filter "Species = Other animals"

We evaluated the search results up through Mar. 2017 and cited the relevant studies. In addition, review papers summarizing past animal studies on nicotine effects were also used as sources to extract relevant studies. However, it should be noted that the search results may not cover all animal models that have been used for studying these diseases, since the keywords used were the name of the disease rather than the name of the animal model.

These searches although likely to identify the majority of the relevant literature are not intended to be absolutely inclusive. Nevertheless, it can be expected that the majority of studies in these areas have been identified and thus the review seems likely to be reasonably representative of the current state of the science in these areas.

Part I

Overview of Epidemiological Data on Smoking and Preclinical and Clinical Data on Nicotine

The first section of this book is dedicated to a high-level overview of currently available epidemiological data regarding the impact of smoking on several neurodegenerative and psychiatric diseases. Available clinical and pre-clinical data regarding the effects of nicotine are summarized for each disease.

Chapter 1

Parkinson's Disease

1.1 SYMPTOMS AND EPIDEMIOLOGY

Parkinson's disease (PD) is a progressive neurodegenerative disorder that affects movement. Early signs may be mild and go unnoticed. Symptoms often begin on one side of the body and usually remain worse on that side, even after both sides have been affected. These motor deficits are due to the loss of dopaminergic neurons in the substantia nigra pars compacta.[39,40] Although the deficit in this brain region is the most severe, other neurotransmitter systems are affected, and these most likely underlie the autonomic problems, cognitive decline, changes in affect, and sleep disturbances.[22]

Approximately 1% of individuals older than 60 years are affected by PD, but younger people can also develop PD. Around 60,000 Americans are diagnosed with PD each year, yet this number does not reflect the thousands of cases that go undetected. It has been estimated that 7–10 million people worldwide are living with PD.[41] The incidence of PD increases with age, but an estimated 4% of PD patients are diagnosed before the age of 50. In the United Kingdom, 1 person in every 500 has PD, with prevalence ranging from 105 to 178 persons with PD per 100,000 population, after adjusting for age.[42]

1.2 MOLECULAR MECHANISMS

The etiology of PD is not completely understood and has been attributed to a complex interplay between genetic and environmental factors. Each of these factors conveys increased risk of disease. A minority of cases (approximately 5%) are genetic (familial), displaying Mendelian inheritance, while the rest are sporadic cases.[43] The pathogenesis of PD involves multiple, related processes including mitochondrial dysfunction, oxidative and nitrative stress, microglial activation, inflammation, aggregation of α-synuclein, and impaired autophagy.[44–46] For many of these factors, the question of whether they are causative or consequential remains open, and they are probably both.

Clinical symptoms appear when around 70% of the dopaminergic neurons in the substantia nigra pars compacta are lost. Clinical, neuropathologic, and neuroimaging findings suggest that the neurodegenerative process of PD begins many years before the onset of motor manifestations. More recent or ongoing studies are largely devoted to the identification of potential markers for the

Nicotine and Other Tobacco Compounds in Neurodegenerative and Psychiatric Diseases.

https://doi.org/10.1016/B978-0-12-812922-7.00001-9

preclinical, or at least premotor stage of the disease, in which dopaminergic neurons are relatively spared and neuroprotective treatments may have a potential effect.[47] Although the precise mechanisms of PD pathogenesis are only partially understood, it is now widely accepted that the accumulation and aggregation of α-synuclein, a principal component of Lewy pathology, plays a crucial role in the pathogenesis of familial and sporadic PD.[48] The precise reason why Lewy bodies develop is not known, and it appears that the Lewy pathology is neither the first sign nor the prerequisite for PD development.[49] Alpha-synuclein may contribute to PD pathogenesis in a number of ways, but it is generally thought that its aberrant soluble oligomeric conformations are the toxic species that mediate disruption of cellular homeostasis and neuronal death, through effects on various intracellular targets, including synaptic function. Furthermore, secreted α-synuclein may exert deleterious effects on neighboring cells, including seeding of aggregation, thus possibly contributing to disease propagation. Accumulated α-synuclein was also found in patients with other neurodegenerative conditions, collectively termed "synucleinopathies."

Based on the literature and current clinical practices, the main therapy for PD is dopamine replacement. This provides effective motor symptom control, particularly in the early disease stage. As the pathology progresses and various nonmotor deficits occur, dopamine replacement does not adequately manage the disease symptoms. In addition, dopamine replacement induces a variety of motor and psychiatric side effects. In May 2006, the Food and Drug Administration (FDA) approved rasagiline to be used with levodopa (L-DOPA) in patients with advanced PD or as a single-drug treatment for early PD. Even very recent approaches, such as deep brain stimulation,[50] provide only symptomatic relief, while the underlying disease continues to worsen. Aside from the importance of symptomatic relief, the shortcomings of current therapies for PD highlight the importance of identifying novel treatment/protection strategies that delay or halt disease progression or ideally restore function.

1.3 IMPACT OF SMOKING, SNUS, AND NICOTINE

1.3.1 Smoking and PD: Epidemiological Evidence

Extensive literature demonstrates reduced PD risk among current and former smokers. This inverse association between PD and smoking correlates with increased intensity and duration of smoking, is more pronounced in current versus former smokers, decreases with years after quitting smoking, and is observed with different types of tobacco products.[51–58] It has been hypothesized that this surprising effect may be due to the increased mortality associated with smoking-related diseases prior to the development of PD. However, several large epidemiological studies have found that this inverse association did not appear to be due to selective survival of PD cases.[59,60] Using a novel modeling method to derive unbiased estimates (the effects of modifying factors are conditioned on total exposure), Van der Mark and colleagues reported a strong inverse association between smoking and PD risk for ever smokers versus the highest quartile of tobacco use.[61] The observed effect of total smoking (all pack years) was significantly modified by time since cessation with a diminishing inverse correlation after smoking cessation. Intensity and duration appear to have an equal contribution on reduced PD risk. Table 1 summarizes meta-analyses with compelling evidence for a reverse correlation between smoking and risk of PD development. Two other clinical studies published subsequently also reported compelling evidence for a reverse correlation between smoking and risk of PD development. In cohort studies, Saaksjarvi et al. and Ton et al. reported OR 0.23, CI 95% (0.08–0.67), and OR 0.42, CI 95% (0.22–0.79), respectively, for ever smokers.[62,63]

1.3.2 Snus and PD: Epidemiological Evidence

A strong inverse association between smokeless tobacco use, including chewing tobacco and snus, and PD risk has been reported. In a prospective cohort of 95,981 never-smoking men, smokeless tobacco use was inversely associated with PD mortality.[64] The authors of this study acknowledged the limitation that the reporting of PD in death certificates was incomplete in 70%–75% of cases. These results are consistent with the findings of Benedetti and colleagues, who reported a stronger reduction in PD incidence among ever users of smokeless tobacco in a case-control study (OR 0.18; 95% CI, 0.04–0.82) than among ever smokers (OR 0.69; 95% CI, 0.45–1.08), although in a much smaller study.[65]

Yang et al. in 2016 reported an analysis of the association between snus use and PD risk, based on seven cohort studies on snus and tobacco smoking in a Swedish male population. The meta-analysis looked at 348,601 men without a history of PD diagnosis at baseline, of whom 107,838 (30.9%) reported ever use of snus, and 165,273 (47.4%) were never tobacco smokers. Nonsmoking men who used snus had a 60% lower risk of PD compared with never snus users. Results also indicated an inverse dose-response relationship between snus use and PD risk, suggesting that nicotine or other components of tobacco leaves may influence the development of PD.[66]

1.3.3 Nicotine and PD: Clinical Evidence

Several studies have demonstrated that nicotine stimulates dopamine release,[67–69] which is the basis of the hypothesis that nicotine may underlie the inverse correlation between smoking and PD. Additionally, the rationale for investigating a role of nicotine was further substantiated by the close anatomical relationship between the nicotinic cholinergic and dopaminergic neurotransmitter systems in the striatum.[70] Results from *in vitro* studies using neuronal cell lines and primary cultures from striatal, nigral, cortical, cerebellar, and other brain regions suggest that nicotine exerts its neuroprotection via $\alpha4\beta2$ subunit-containing or $\alpha7$ nAChRs.[25,71–73]

TABLE 1 Epidemiological Studies on Smoking and Parkinson's Disease—Meta-Analyses

Studies Included	Factors	OR[a]/RR[b] (95% CI[c])	References
25 Case-control studies and 8 cohort studies between 2000 and 2013	Current smoker	0.41 (0.34–0.48)	1
61 Case-control and 8 cohort studies between 1959 and 2014	Ever smoker	0.59 (0.56–0.62)	2
61 Case-control and 6 cohort studies between 1966 and 2011	Ever smoker	0.64 (0.60–0.69)	3
26 Case-control and 7 cohort studies between 1966 and 2011	Current smoker	0.44 (0.39–0.50)	
48 Case-control and 6 cohort studies before 2010	Ever smoker	0.55 (0.51–0.59)	4
	Current smoker	0.31 (0.25–0.38)	
6 Cohort studies between 1959 and 1997	Ever smoker	0.53 (0.42–0.66)	5
	Current smoker	0.37 (0.24–0.58)	
	Past smoker	0.63 (0.41–0.96)	
44 Case-control and 4 cohort studies between 1968 and 2001	Ever smoker	0.59 (0.54–0.63)	6
	Current smoker	0.39 (0.32–0.47)	
	Past smoker	0.80 (0.69–0.93)	
40 Case-control and 6 cohort studies before 2000	Ever smoker	0.57 (0.52–0.63)	7

[a] *OR, odds ratio.*
[b] *RR, relative risk.*
[c] *CI, confidence interval.*

1. Breckenridge CB, Berry C, Chang ET, Sielken Jr RL, Mandel JS. Association between Parkinson's disease and cigarette smoking, rural living, well-water consumption, farming and pesticide use: systematic review and meta-analysis. *PLoS One* 2016;**11**:e0151841.
2. Li X, Li W, Liu G, Shen X, Tang Y. Association between cigarette smoking and Parkinson's disease: a meta-analysis. *Arch Gerontol Geriatr* 2015;**61**:510–6.
3. Noyce AJ, Bestwick JP, Silveira-Moriyama L, et al. Meta-analysis of early nonmotor features and risk factors for Parkinson disease. *Ann Neurol* 2012;**72**:893–901.
4. Kiyohara C, Kusuhara S. Cigarette smoking and Parkinson's disease: a meta-analysis. *Fukuoka Igaku Zasshi=Hukuoka Acta Med* 2011;**102**:254–65.
5. Allam MF, Campbell MJ, Hofman A, Del Castillo AS, Fernandez-Crehuet Navajas R. Smoking and Parkinson's disease: systematic review of prospective studies. *Mov Disord* 2004;**19**:614–21.
6. Hernan MA, Takkouche B, Caamano-Isorna F, Gestal-Otero JJ. A meta-analysis of coffee drinking, cigarette smoking, and the risk of Parkinson's disease. *Ann Neurol* 2002;**52**:276–84.
7. Sugita M, Izuno T, Tatemichi M, Otahara Y. Meta-analysis for epidemiologic studies on the relationship between smoking and Parkinson's disease. *J Epidemiol* 2001;**11**:87–94.

Clinical trials using nicotine in PD patients have been conducted (Table 2). The published results are not conclusive with regard to the effect of nicotine on clinical symptoms: Two studies reported positive effects, and two others reported no effects. It has been proposed that the reason for these differential outcomes may relate to variations in the mode of nicotine administration (patch, gum, and intravenous); inadequate dose, timing, or duration (days to weeks) of treatment and differences in the degree of parkinsonism and type of trial (open-label vs. double-blinded). Nevertheless, the fact that several PD symptoms improved (e.g., less muscular rigidity and fewer tremors) indicates that nicotine should be investigated further in this context. Several ongoing trials are noted in ClinicalTrials.gov although the outcome is not yet available.

1.3.4 Nicotine and PD: Preclinical Evidence

A number of studies using experimental animal models (Table 3) have shown that nicotine can protect against neurotoxin-induced nigrostriatal damage and improve motor impairments associated with L-DOPA (the gold standard therapy for PD), with dyskinesia as a side effect.[74] In rats, nicotine pretreatment before introduction of lesions reduces neuronal damage, assessed using markers of striatal dopaminergic integrity, including levels of dopamine and its metabolites, tyrosine hydroxylase, dopamine transporters, and vesicular monoamine transporters. The degree of protection against nigrostriatal damage depends on several parameters: lesion size (optimal effectiveness is observed when a moderate-damage regimen is engaged), nicotine dose, and timing of administration. Nicotine exhibits a U-shaped dose-response curve. Maximal protection occurs with intermediate nicotine-dosing regimens. Protection was observed with both intermittent and continuous nicotine dosing. Different animal PD models in studies of nicotine are reviewed in more details by Quik et al.[74]

1.3.5 nAChRs in PD

Several studies have analyzed the decline in specific nAChRs in PD patients. See Section 9.1.2 for details about the pharmacology of nAChRs. Court and colleagues reported the decline of α3 subunits and no change in α7 subunits.[75] This observation was confirmed by another study, where selective losses of the α3- and β2-containing nAChRs and an increase in the α7 nAChRs were reported.[76] There appears to be a particularly pronounced decrease in the subpopulation of α6β2 nAChRs, suggesting a critical role for α6-containing receptors in PD pathology.[77] The decrease in the α4β2 nAChR population that does not include α6 is present, but much less severe.[43]

Roles for nAChRs as potential targets for α-synuclein-induced neurotoxicity resulting in neuronal degeneration are possible, but have not been extensively investigated. By using a pharmacological approach in SH-EP1 cell cultures expressing transfected human nAChRs, a clear inhibitory effect of oligomeric

TABLE 2 Clinical Trials on Nicotine and PD[a]

Type of Study	Subject Description	Nicotine Treatment	Outcome	References
Placebo-controlled	Patient 1, a 69-year-old female (never smoker) Patient 2, a 64-year-old male (former smoker)	Nicotine gum and/or patch	Patient 1—decreased tremors, less muscular rigidity, reduced disorganized thinking, and improved sleep Patient 2—diminished bradykinesia and increased energy	1
Randomized, double-blind, placebo-controlled	48 Patients (nonsmokers)	25 Patients received 2 mg nicotine gum; three pieces taken at 2 h intervals within 4.5 h. 23 Patients received a placebo	No significant differences in parkinsonian symptoms between treatment groups	2
Open-label	15 Early-to-moderate patients	Acute intravenous nicotine up to 1.25 mg/kg/min and transdermal nicotine up to 14 mg/day for 2 weeks	Improved cognitive performance, but not attention and semantic retrieval after acute injection; sustained (up to 1 month) improvement in motor function after chronic dose	3
Randomized, double-blind, placebo-controlled	32 Patients (nonsmokers for at least 2 years)	16 Patients received a nicotine patch (7 or 14 mg/day) for 2 weeks. 16 patients received a placebo patch for 2 weeks	No significant drug effect at either dose	4
Open- label	22 Patients (nonsmokers)	Increasing transdermal nicotine over 25 days up to 21 mg/day	Poor nicotine tolerance (13 withdrew due to acute side effects). No effects on motor or cognitive functions in remaining 9 patients	5

Pilot, open-label study	6 Idiopathic patients (never smokers)	Increasing daily dose of transdermal nicotine up to 105 mg/day from week 1 to week 14; maintained the highest dose for 3 weeks	Improved motor score and reduced dopaminergic treatment; frequent but moderate nausea and vomiting	6
Open-label	6 Patients (never smokers)	Patients received transdermal nicotine and were followed up for 29 weeks	All patients improved motor scores at 3 months, and most received fewer dopaminergic drugs. Motor improvement persisted to a lesser extent at 1 year	7
Double-blind, placebo-controlled, crossover	12 Patients and 17 controls (all nonsmokers)	Participants received nicotine patch (7 mg/day) or placebo in the first session, then the alternative in the second session, 7–10 days later	Improved compromised semantic processing in PD	8

[a]*PD*, Parkinson's disease.

1. Fagerström KO, Pomerleau O, Giordani B, Stelson F. Nicotine may relieve symptoms of Parkinson's disease. *Psychopharmacology* 1994;**116**:117–9.
2. Clemens P, Baron J, Coffey D, Reeves A. The short-term effect of nicotine chewing gum in patients with Parkinson's disease. *Psychopharmacology* 1995;**117**:253–6.
3. Kelton MC, Kahn HJ, Conrath CL, Newhouse PA. The effects of nicotine on Parkinson's disease. *Brain Cogn* 2000;**43**:274–82.
4. Vieregge A, Sieberer M, Jacobs H, Hagenah J, Vieregge P. Transdermal nicotine in PD A randomized, double-blind, placebo-controlled study. *Neurology* 2001;**57**:1032–5.
5. Lemay S, Chouinard S, Blanchet P, et al. Lack of efficacy of a nicotine transdermal treatment on motor and cognitive deficits in Parkinson's disease. *Prog Neuropsychopharmacol Biol Psychiatry* 2004;**28**:31–9.
6. Villafane G, Cesaro P, Rialland A, et al. Chronic high dose transdermal nicotine in Parkinson's disease: an open trial. *Eur J Neurol* 2007;**14**:1313–6.
7. Itti E, Villafane G, Malek Z, et al. Dopamine transporter imaging under high-dose transdermal nicotine therapy in Parkinson's disease: an observational study. *Nucl Med Commun* 2009;**30**:513–8.
8. Holmes AD, Copland DA, Silburn PA, Chenery HJ. Acute nicotine enhances strategy-based semantic processing in Parkinson's disease. *Int J Neuropsychopharmacol* 2011;**14**:877–85.

TABLE 3 Nicotine Treatment in PD[a]—Animal Studies

Nicotine Administration	Animal Species	Chemical Used to Induce Substantia Nigra Lesion	Conclusion	References
s.c.[b] (1 mg/kg), pre- and postlesion	Rat	6-OHDA[c]	Pretreatment reduced damage	1
Rat, s.c. (0.75, 1.5, 3.0, and 30.0 mg/kg/day for 14 days); prelesion Mouse, s.c. (1 mg/kg, acute); prelesion	Rat Mouse α4 nAChR knockout	Rat, 6-OHDA Mouse, methamphetamine	Pretreatment reduced damage in rats and wild-type mice, but not α4 nAChR[d] knockout mice	2
i.p.[e] (0.5 mg/kg), immediately after lesioning and every 30 min, followed by minipump use (0.125 mg/kg/h)	Rat	Hemitransection of the mesodiencephalic junction	Protection in substantia nigra and striatum	3
s.c. (0.6 or 0.8 mg/kg), daily for 12 days postlesioning	Rat	Medial forebrain bundle 6-OHDA (6 mg)	Protection in substantia nigra but not striatum	4
Increasing doses in drinking water (up to 650 μg/mL) and food pellets, pre- and postlesion	Adult female squirrel monkey (*Saimiri sciureus*)	MPTP[f]	Attenuated the effects of neuronal damage	5
Chronic nicotine injection (2 mg/kg, s.c., four doses/day)	Mouse	MPTP	Antagonized MPTP-induced dopamine depletion in the substantia nigra Protected against MPTP-suppressed spontaneous locomotor activity	6
Chronic nicotine treatment i.p. (1 mg/kg) every 2 h for 4 weeks	Rat	Lipopolysaccharide-induced TH[g]-positive neuronal loss in the substantia nigra	Protected against TH nigral neuronal loss Reduced TNF-α formation	7

Drinking water (30 mg/L) over 6 weeks	Hemiparkinson rat	6-OHDA	Reduced amphetamine-induced circling behavior by 40% Modified dopamine receptor dynamics	8
Drinking water (300 µg/mL) over 6 months	Monkey	L-DOPA-induced dyskinesia	Decreased L-DOPA-induced dyskinesia by approximately 70%	9,10
i.p. (0.2 or 2 mg/kg), five times/day at 2 h intervals. High- and low-exposure groups for cigarette smoke	C57BL/6 mouse	MPTP (four injections of 15 mg/kg in 1 day at 2 h intervals prior to nicotine/smoke administration)	Reduced loss of TH and neurons attributed to nicotine and low cigarette smoke exposure	11

[a] *PD, Parkinson's disease.*
[b] *s.c., subcutaneous.*
[c] *6-OHDA, 6-hydroxydopamine.*
[d] *nAChR, nicotinic acetylcholine receptor.*
[e] *i.p., intraperitoneal.*
[f] *MPTP, 1-methyl-4-phenyl-1,2,3,6-tetrahydropyridine.*
[g] *TH, tyrosine hydroxylase.*

1. Costa G, Abin-Carriquiry J, Dajas F. Nicotine prevents striatal dopamine loss produced by 6-hydroxydopamine lesion in the substantia nigra. *Brain Res* 2001;**888**:336–42.
2. Ryan R, Ross S, Drago J, Loiacono R. Dose-related neuroprotective effects of chronic nicotine in 6-hydroxydopamine treated rats, and loss of neuroprotection in α4 nicotinic receptor subunit knockout mice. *Br J Pharmacol* 2001;**132**:1650–6.
3. Janson A, Møller A. Chronic nicotine treatment counteracts nigral cell loss induced by a partial mesodiencephalic hemitransection: an analysis of the total number and mean volume of neurons and glia in substantia nigra of the male rat. *Neuroscience* 1993;**57**:931–41.
4. Visanji N, O'Neill M, Duty S. Nicotine, but neither the α4β2 ligand RJR2403 nor an α7 nAChR subtype selective agonist, protects against a partial 6-hydroxydopamine lesion of the rat median forebrain bundle. *Neuropharmacology* 2006;**51**:506–16.
5. Quik M, Chen L, Parameswaran N, Xie X, Langston JW, McCallum SE. Chronic oral nicotine normalizes dopaminergic function and synaptic plasticity in 1-methyl-4-phenyl-1,2,3,6-tetrahydropyridine-lesioned primates. *J Neurosci* 2006;**26**:4681–9.
6. Gao Z, Cui W, Zhang H, Liu C. Effects of nicotine on 1-methyl-4-phenyl-1,2,5,6-tetrahydropyridine-induced depression of striatal dopamine content and spontaneous locomotor activity in C57 black mice. *Pharmacol Res* 1998;**38**:101–6.
7. Park HJ, Lee PH, Ahn YW, et al. Neuroprotective effect of nicotine on dopaminergic neurons by anti-inflammatory action. *Eur J Neurosci* 2007;**26**:79–89.
8. García-Montes J-R, Boronat-Garcia A, Lopez-Colome A-M, Bargas J, Guerra-Crespo M, Drucker-Colin R. Is nicotine protective against Parkinson's disease? An experimental analysis. *CNS Neurol Disord Drug Targets* 2012;**11**:897–906.
9. Quik M, Mallela A, Ly J, Zhang D. Nicotine reduces established levodopa-induced dyskinesias in a monkey model of Parkinson's disease. *Mov Disord* 2013;**28**:1398–406.
10. Zhang D, Bordia T, McGregor M, McIntosh JM, Decker MW, Quik M. ABT-089 and ABT-894 reduce levodopa-induced dyskinesias in a monkey model of Parkinson's disease. *Mov Disord* 2014;**29**:508–17.
11. Parain K, Hapdey C, Rousselet E, Marchand V, Dumery B, Hirsch EC. Cigarette smoke and nicotine protect dopaminergic neurons against the 1-methyl-4-phenyl-1,2,3,6-tetrahydropyridine parkinsonian toxin. *Brain Res* 2003;**984**:224–32.

α-synuclein on α4β2 function was recorded using electrophysiological measurements of currents, but no effect on α4β4 or α7 nAChRs was shown. In addition, only large oligomeric α-synuclein aggregates exhibited the inhibitory effect on human α4β2 nAChRs.[78]

1.3.6 Summary

Epidemiological data on smoking and snus consumption (in a Swedish population) and the incidence of PD have clearly shown an inverse correlation. Nicotine was a topic of preclinical and clinical investigations into its therapeutic potential because nicotine stimulates dopamine release, a property relevant to PD. Clinical data were inconclusive, with some positive and some negative results following administration of nicotine via different routes (nicotine patch, gum, or intravenous). Given the roles of different nAChRs in PD, nicotine should be investigated further for its pharmacological properties. The inconsistency between epidemiological data and clinical trial results obtained with nicotine alone may suggest that other tobacco compounds, separated from combustible tobacco products, may play a role.

Chapter 2

Alzheimer's Disease

2.1 SYMPTOMS AND EPIDEMIOLOGY

Alzheimer's disease (AD) is the most common cause of dementia, a group of brain disorders that cause the loss of intellectual and social skills. The most common early symptom is difficulty in remembering recent events (short-term memory loss). As the disease advances, symptoms include problems with language, disorientation (including easily getting lost), mood swings, loss of motivation, not managing self-care, and behavioral issues such as aggression. As patients' condition declines, they often withdraw from family and society. Gradually, bodily functions are lost, ultimately leading to death. Brain cells degenerate and die in AD, causing a steady decline in memory and mental function. The key pathological changes observed in AD brain tissue are extracellular deposition of amyloid β (Aβ) peptides in diffuse and neuritic plaques and hyperphosphorylated tau protein, a microtubule assembly protein that accumulates intracellularly as neurofibrillary tangles (NFTs). Additional changes include reactive microgliosis and widespread loss of neurons, white matter, and synapses.

Worldwide, nearly 44 million people suffer from AD or a related dementia. According to AD international, most people currently living with dementia have not received a formal diagnosis, and only one in four people with AD has been diagnosed.[79] There are two types of AD, based on age of onset: Early-onset AD (onset <65 years) accounts for 1%–5% of all cases, and late-onset AD (onset ≥ 65 years) accounts for $>95\%$ of affected cases.

2.2 MOLECULAR MECHANISMS

The exact mechanisms leading to AD development are yet to be determined. Causes of AD likely include a combination of genetic, environmental, and lifestyle factors. Three genes, amyloid precursor protein (*APP*) and presenilin 1 and 2 (*PSEN1* and *PSEN2*), encode proteins involved in amyloid precursor breakdown and Aβ generation and are implicated in the pathophysiology of early-onset AD. This form of AD is generally associated with a more rapid rate of progression and a Mendelian pattern of inheritance. The apolipoprotein E (*APOE*) gene is involved in late-onset AD. *APOE* ε4 is the isoform associated with increased risk of AD development and is also associated with an earlier age of disease onset. However, carrying *APOE* ε4 does not necessarily mean that a

Nicotine and Other Tobacco Compounds in Neurodegenerative and Psychiatric Diseases.

https://doi.org/10.1016/B978-0-12-812922-7.00002-0

person would definitely develop AD, and some people with no *APOE* ε4 alleles may also develop the disease. Genome-wide association studies have revealed a number of other genes and genomic regions involved in AD.[80] In addition, late-onset AD arises from a complex series of brain changes that occur over decades.[81] Health-related risk factors for AD development are vascular conditions such as heart disease, stroke, and high blood pressure and metabolic conditions such as diabetes, obesity, and traumatic brain injury. Diet, physical activity, and intellectual pursuits are lifestyle parameters that can influence the risk for AD development in both directions.

2.3 IMPACT OF SMOKING AND NICOTINE

2.3.1 Smoking and AD: Epidemiological Evidence

Epidemiological studies on AD and smoking have provided mixed results on how smoking impacts the risk of AD development. Two case-control studies from the 1990s suggested that smoking lowers AD risk,[82,83] whereas prospective studies indicate that smoking increases this risk[17,84–86] or has no effect on the probability of developing AD.[87,88] It has been suggested that tobacco industry affiliation has an influence on the outcome of such studies, and indeed, a series of meta-analyses have been performed in an attempt to evaluate whether there are differences between tobacco-industry-affiliated studies and those that have no affiliation with the tobacco industry. Without further comment on this possibility, it can be seen in Table 1 that the results of these meta-analyses generally suggest an odds ratio around or above parity, with only one meta-analysis indicating a statistically significant protective effect. Smoking is a recognized risk factor for development of AD.[8] Smoking may affect AD risk via several

TABLE 1 Epidemiological Studies on Smoking and AD[a]—Meta-Analyses

Studies Included	Factors	OR[b]/RR[c] (95% CI[d])	References
22 Cohort studies before 2014	Ever smoker	1.12 (1.00–1.26)	1
12 Cohort studies before 2014	Current smoker	1.40 (1.13–1.73)	
13 Cohort studies before 2014	Past smoker	1.04 (0.96–1.13)	
9 Cohort studies between 1990 and 2012	Ever smoker	1.37 (1.23–1.52)	2
4 Studies before 2014	Ever smoker	1.55 (1.15–1.95)	3
6 Studies before 2014	Current smoker	1.52 (1.18–1.86)	
6 Studies before 2014	Past smoker	0.94 (0.76–1.12)	

TABLE 1 Epidemiological Studies on Smoking and AD—Meta-Analyses—cont'd

Studies Included	Factors	OR/RR (95% CI)	References
18 Case-control studies without tobacco industry affiliation	Ever smoker	0.91 (0.75–1.10)	4
8 Case-control studies with tobacco industry affiliation	Ever smoker	0.86 (0.75–0.98)	
14 Cohort studies without tobacco industry affiliation	Ever smoker	1.45 (1.16–1.80)	
3 Cohort studies with tobacco industry affiliation	Ever smoker	0.60 (0.27–1.32)	
8 Cohort studies between 1996 and 2007	Current smoker	1.59 (1.15–2.20)	5
Not specified	Past smoker	0.99 (0.81–1.23)	
3 Cohort studies before 2005	Ever smoker	1.21 (0.66–2.22)	6
4 Cohort studies before 2005	Current smoker	1.79 (1.43–2.23)	
5 Cohort studies before 2005	Past smoker	1.01 (0.83–1.23)	
21 Case-control studies	Ever smoker	0.74 (0.66–0.84)	7
8 Cohort studies before 2000	Ever smoker	1.10 (0.94–1.29)	

[a] *AD, Alzheimer's disease.*
[b] *OR, odds ratio.*
[c] *RR, relative risk.*
[d] *CI, confidence interval.*

1. Zhong G, Wang Y, Zhang Y, Guo JJ, Zhao Y. Smoking is associated with an increased risk of dementia: a meta-analysis of prospective cohort studies with investigation of potential effect modifiers. *PLoS One* 2015;**10**:e0118333.
2. Beydoun MA, Beydoun HA, Gamaldo AA, Teel A, Zonderman AB, Wang Y. Epidemiologic studies of modifiable factors associated with cognition and dementia: systematic review and meta-analysis. *BMC Public Health* 2014;**14**:1.
3. Prince M, Albanese E, Guerchet M, Prina M. *World Alzheimer Report 2014. Dementia and risk reduction: an analysis of protective and modifiable factors*. Londres: Alzheimers Disease International; 2014.
4. Cataldo JK, Prochaska JJ, Glantz SA. Cigarette smoking is a risk factor for Alzheimer's disease: an analysis controlling for tobacco industry affiliation. *J Alzheimer's Dis* 2010;**19**:465–80.
5. Peters R, Poulter R, Warner J, Beckett N, Burch L, Bulpitt C. Smoking, dementia and cognitive decline in the elderly, a systematic review. *BMC Geriatr* 2008;**8**:36.
6. Anstey KJ, von Sanden C, Salim A, O'Kearney R. Smoking as a risk factor for dementia and cognitive decline: a meta-analysis of prospective studies. *Am J Epidemiol* 2007;**166**:367–78.
7. Almeida OP, Hulse GK, Lawrence D, Flicker L. Smoking as a risk factor for Alzheimer's disease: contrasting evidence from a systematic review of case–control and cohort studies. *Addiction* 2002;**97**:15–28.

mechanisms: it may increase generation of free radicals, leading to greater oxidative stress, or affect the inflammatory immune system, leading to activation of phagocytes and further oxidative damage.[89] Contradictory findings on the impact of smoking on AD are also evident in more recent meta-analysis (Table 1). In a meta-analysis, Zhong et al. searched different databases and included a total of 37 studies. Smokers showed increased risk of dementia, and smoking cessation decreased the risk to that of never smokers. Increased AD risk from smoking is more pronounced in *APOE* ε4 noncarriers. Survival bias and competing risk reduced the risk of dementia from smoking at an extreme age.[90] Prince et al. reported that after quitting smoking, the positive association between smoking and AD is reversed.[91]

2.3.2 Nicotine and AD: Clinical Evidence

In AD patients, acetylcholine (ACh) release in the brain is dramatically reduced because of cholinergic neuronal loss.[92,93] The number of α7 nAChRs is reduced but upregulated by chronic nicotine treatment, which is the basis for the hypothesis that nicotine-only treatment may be beneficial for AD patients. Acute nicotine administration improves some symptoms of AD, such as recall, visual attention, and mood. To date, several clinical trials have been conducted with nicotine administered to AD patients through transdermal patches (Table 2). Certain symptoms were clearly improved, such as sustained improvement in attention and verbal learning. On the other hand, other studies found no significant effects of nicotine on memory performance in healthy and AD patients.[94a,255] Critiques included the suboptimal quality of the clinical trials (no double-blind, placebo-controlled, randomized settings) on nicotine in AD patients, indicating the lack of evidence to recommend nicotine as a treatment for AD.[94] It has been shown that chronic nicotine exposure can lead to loss of nicotinic functional activity as a result of the persistent deactivation of nAChR receptors (i.e., nAChRs desensitization), a mechanism that might, in part, explain the modest or lack of effect observed in AD patients.[94b]

Several clinical trials have been conducted in patients with mild cognitive impairment (MCI), which is defined as a subjective and objective decline in cognition and function that does not meet the criteria for a diagnosis of dementia. Mild cognitive impairment represents a transitional state between cognition of normal aging and mild dementia, where the changes are not severe enough to interfere with daily life or independent function. Those with MCI show increased risk of eventually developing AD or another type of dementia. However, not all MCI cases become worse, and some eventually improve. Newhouse et al. conducted a study on nonsmokers with MCI (67 subjects completed, 34 nicotine and 33 placebo) and demonstrated that transdermal nicotine (15 mg/day) can be safely administered to nonsmoking subjects with MCI over 6 months with improvement in primary and secondary cognitive measures of attention, memory, and mental processing, but not in ratings of clinician-rated global impression.[95]

TABLE 2 Clinical Trials on Nicotine and AD[a]

Type of Study	Subject Description	Nicotine Treatment	Outcome	References
Placebo-controlled	6 Patients (all nonsmokers)	Increasing doses of nicotine base and placebo (0.125, 0.25, and 0.5 mg/kg/min, intravenous infusion) Each dose given on a separate day, 48 h apart	Decrease in intrusion errors Significant dose-related increase in anxiety and depressive effects	1
Single-blind, placebo-controlled	22 Patients, 24 young healthy controls, and 24 aged healthy controls (with equal numbers of smokers and nonsmokers in each group)	Each subject given a placebo followed by increasing doses of nicotine (0.4, 0.6, and 0.8 mg, s.c.[b]), followed by placebo	Acute nicotine significantly improved perception and sustained visual attention, rapid visual information processing, and reaction time	2
Double-blind, placebo-controlled, crossover	8 Patients (all nonsmokers)	First session—nicotine patch (5 mg/kg for 1 week, then 10 mg/kg for 2 weeks, and followed by 5 mg/kg for 1 week) Second session—placebo patch for same duration	Sustained improvement in attention No effect on cognition and other behavioral domains	3
Double-blind, placebo-controlled	102 Patients	42 Subjects received a 7 mg/kg nicotine patch for 6 weeks 35 Subjects received a 14 mg/kg nicotine patch for 6 weeks 25 Subjects received a control patch	Significant effect on verbal learning, objective learning, delayed recall, and word retrieval No effect on measures of concentration or psychomotor speed	4
Single-blind, placebo-controlled, crossover	15 Healthy subjects	First session—nicotine patch (7 mg/kg) or placebo Second session—alternative patch	Significantly enhanced regional efficiency in limbic and paralimbic areas, which are altered in AD and schizophrenia	5

[a] *AD, Alzheimer's disease.*
[b] *s.c., subcutaneous.*

1. Newhouse PA, Sunderland T, Tariot PN, et al. Intravenous nicotine in Alzheimer's disease: a pilot study. *Psychopharmacology* 1988;**95**:171–5.
2. Jones G, Sahakian B, Levy R, Warburton DM, Gray J. Effects of acute subcutaneous nicotine on attention, information processing and short-term memory in Alzheimer's disease. *Psychopharmacology* 1992;**108**:485–94.
3. White HK, Levin ED. Four-week nicotine skin patch treatment effects on cognitive performance in Alzheimer's disease. *Psychopharmacology* 1999;**143**:158–65.
4. Howe MN, Price IR. Effects of transdermal nicotine on learning, memory, verbal fluency, concentration, and general health in a healthy sample at risk for dementia. *Int Psychogeriatr* 2001;**13**:465–75.
5. Wylie KP, Rojas DC, Tanabe J, Martin LF, Tregellas JR. Nicotine increases brain functional network efficiency. *Neuroimage* 2012;**63**:73–80.

The authors commented that larger clinical studies are required to confirm clinical relevance. A new trial is underway with support from the National Institute of Aging and the Alzheimer's Drug Discovery Foundation to more effectively test the effects of transdermal nicotine patches for patients with mild cognitive impairment.

2.3.3 Nicotine and AD: Preclinical Evidence

Both positive and negative effects of nicotine on AD markers have been observed *in vivo*. Procognitive effects of nicotine have been demonstrated in rodents and nonhuman primates,[96,97] with neuroprotective activity in a variety of model systems. Noshita and colleagues recently published results from nicotine tested in an Aβ1-42-induced animal model of AD. In this study, nicotine administered intraperitoneally to rats (0.2 mg/kg, once a day for 9 weeks beginning 3 weeks after the Aβ infusion) ameliorated learning and memory deficits. This improvement was mediated, at least in part, by enhancement of cholinergic neurotransmission.[98] Another study, published in 2010 by Deng et al., showed that nicotine (1 mg/kg) treatment significantly exacerbates cognitive impairment and tau phosphorylation in the hippocampus compared with the effects in Aβ 25–35-only injection groups in the Aβ rat model of AD. Other preclinical results with mixed effects of nicotine in AD models are summarized in Table 3.

2.3.4 nAChRs in AD

Progressive loss of nAChRs in the cerebral cortex and hippocampus is marked in AD patients and is associated with cognitive decline.[99,100] Positron emission tomography scanning suggests that a deficit in nAChRs is an early event in AD progression. nAChR protein content determination has yielded somewhat inconsistent results. Reduced levels of the α4, α3, and α7 nAChR subtypes in hippocampi of AD brains were reported by Guan et al.,[101] while no change in α3 and α7 in same regions was reported in another study.[102] Differences at the transcriptional levels of various nAChR subunits did not correspond to protein level modifications observed in AD patients. Such discrepancies may be explained by posttranslational modifications and impact of the overall pathological state on receptor turnover.[103]

One of the pathological hallmarks of AD is accumulation of Aβ protein in the brain. Aβ aggregation in various forms, ranging from dimers and oligomers to fibrils in amyloid plaques (but not as a monomer), causes neurodegeneration.[104,105] Interaction between Aβ and nAChRs is a possible explanation for the cholinergic systems dysfunction that is a characteristic of AD. Several studies investigated and reported interaction between Aβ and nAChRs and activation of signaling cascades that resulted in the disruption of synaptic functions. Different experimental conditions and systems used showed both agonistic and antagonistic effects of Aβ and nAChR interactions. For example, in rat

TABLE 3 Nicotine in AD[a]—Preclinical Studies

Model	Nicotine Treatment	Effect	References
ICV[b]-injected Aβ[c]-induced AD in rat	i.p.[d] 0.2 mg/kg for 9 weeks	Improved learning and memory deficits in water maze test No effect on deficits in passive avoidance test	1
ICV-injected Aβ-induced AD in rat	i.p. 2 mg/kg for 6 weeks	Reduced Aβ and BACE1 in hippocampal CA1 region. Prevention of learning and short-term memory deficits in the radial arm water maze test. Prevention of inhibition of basal synaptic transmission and LTP[e] in CA1	2
Hippocampal Aβ-injected induced cognitive impairments in rat	s.c.[f] 1 mg/kg for 2 weeks	Exacerbated cognitive impairment in Morris water maze test increased tau phosphorylation	3
Hippocampus-infused Aβ-induced AD in mouse	s.c. 0.125 mg/kg for 1 month	Improved spatial learning deficits in Morris water maze test	4
ICV-injected colchicine-induced AD in rat	i.p. 0.5 mg/kg	Improved deficits in Morris water maze test	5
ICV-injected sodium metavanadate-induced learning deficits in rat	Bilateral intrahippocampal infusion of nicotine (1 μg/side)	Diminished learning deficits in Morris water maze test	6
Sleep deprivation-induced learning and memory deficits in rat	s.c. 1 mg/kg, concurrent with sleep deprivation	Attenuated learning impairments and short-term memory deficits in radial arm and Morris water maze tests	7
New intruder-induced chronic psychosocial stress rat model	s.c. 1 mg/kg twice a day for 4–5 weeks	Normalized stress-induced learning and memory deficits in radial arm and water maze tests, which were blocked by nicotinic receptor antagonists	8
APP/PS1 transgenic (APPswe/PS1DE9) mice	i.p. 1 mg/kg, 3 times/week for 1 month	Improved spatial learning in both young (6 months) and old (13 months) mice Synaptic loss prevented	9
APP/PS1 transgenic (APPswe/PS1DE9) mice	Cigarette smoke (1 h/day for 4 months)	Increased amyloid deposition, exacerbated gliosis and neuroinflammation, and induced phosphorylated tau	10

Continued

TABLE 3 Nicotine in AD—Preclinical Studies—cont'd

Model	Nicotine Treatment	Effect	References
Tg2576 (APPswe) mice	Increasing dose (from 0.25 to 0.45 mg/kg) s.c. for 10 days	Decreased insoluble Aβ40 by 77% and insoluble Aβ42 by 84%	11
Tg2576 (APPswe) mice	200 mg/mL in drinking water for 5.5 months	Decreased insoluble Aβ40 by 37% and insoluble Aβ42 by 56%	12
3xTg (APPswe/tauP301L/PS1M145V) mice	600 mg/mL in drinking water for 5 months	No effect on APP processing and Ab Increased tau phosphorylation and aggregation	13
Scopolamine-induced amnesia in rat	*S* (−) nicotine 0.6 and 6 μmol/kg s.c. 20 min before behavioral tests	Acute effect on cognition only. No other pathological hallmarks	14
hAChE[g]-Tg/APPswe 14 months old mice	L (−) and D (+) 0.45 mg/kg twice daily for 10 days	Different effects of L and D nicotine depending on the interaction of Aβ synaptic AChE[h]	15
Tg2576 (APPswe) mice	(−) Nicotine 0.25–0.45 mg/kg up to 10 days s.c.	Reduction in levels of insoluble Aβ 1–40 and Aβ 1–42	16

[a] *AD, Alzheimer's disease.*
[b] *ICV, intracerebroventricular.*
[c] *Aβ,.amyloid β*
[d] *i.p., intraperitoneal.*
[e] *LTP, long-term potentiation.*
[f] *s.c., subcutaneous.*
[g] *hAChE, human acetylcholinesterase.*
[h] *AChE, acetylcholinesterase.*

1. Noshita T, Murayama N, Nakamura S. Effect of nicotine on neuronal dysfunction induced by intracerebroventricular infusion of amyloid-β peptide in rats. *Eur Rev Med Pharmacol Sci* 2015;**19**:334–43.
2. Srivareerat M, Tran TT, Salim S, Aleisa AM, Alkadhi KA. Chronic nicotine restores normal Aβ levels and prevents short-term memory and E-LTP impairment in Aβ rat model of Alzheimer's disease. *Neurobiol Aging* 2011;**32**:834–44.
3. Deng J, Shen C, Wang Y-J, et al. Nicotine exacerbates tau phosphorylation and cognitive impairment induced by amyloid-beta 25–35 in rats. *Eur J Pharmacol* 2010;**637**:83–8.
4. Xue MQ, Liu XX, Zhang YL, Gao FG. Nicotine exerts neuroprotective effects against β-amyloid-induced neurotoxicity in SH-SY5Y cells through the Erk1/2-p38-JNK-dependent signaling pathway. *Int J Mol Med* 2014;**33**:925–33.
5. Rangani RJ, Upadhya MA, Nakhate KT, Kokare DM, Subhedar NK. Nicotine evoked improvement in learning and memory is mediated through NPY Y1 receptors in rat model of 6 Alzheimer's disease. *Peptides* 2012;**33**:317–28.
6. Azami K, Tabrizian K, Hosseini R, et al. Nicotine attenuates spatial learning deficits induced by sodium metavanadate. *Neurotoxicology* 2012;**33**:44–52.
7. Aleisa A, Helal G, Alhaider I, et al. Acute nicotine treatment prevents rem sleep deprivation-induced learning and memory impairment in rat. *Hippocampus* 2011;**21**:899–909.
8. Alzoubi KH, Srivareerat M, Tran TT, Alkadhi KA. Role of α7-and α4β2-nAChRs in the neuroprotective effect of nicotine in stress-induced impairment of hippocampus-dependent memory. *Int J Neuropsychopharmacol* 2013;**16**:1105–13.
9. Inestrosa NC, Godoy JA, Vargas JY, et al. Nicotine prevents synaptic impairment induced by amyloid-β oligomers through α7-nicotinic acetylcholine receptor activation. *Neuromolecular Med* 2013;**15**:549–69.
10. Moreno-Gonzalez I, Estrada LD, Sanchez-Mejias E, Soto C. Smoking exacerbates amyloid pathology in a mouse model of Alzheimer's disease. *Nat Commun* 2013;**4**:1495.
11. Court J, Keverne J, Svedberg M, et al. Nicotine reduces Aβ in the brain and cerebral vessels of APPsw mice. *Eur J Neurosci* 2004;**19**:2703–10.
12. Nordberg A, Hellström-Lindahl E, Lee M, et al. Chronic nicotine treatment reduces β-amyloidosis in the brain of a mouse model of Alzheimer's disease (APPsw). *J Neurochem* 2002;**81**:655–8.
13. Oddo S, Caccamo A, Green KN, et al. Chronic nicotine administration exacerbates tau pathology in a transgenic model of Alzheimer's disease. *Proc Natl Acad Sci U S A* 2005;**102**:3046–51.
14. Lippiello P, Bencherif M, Gray J, et al. RJR-2403: a nicotinic agonist with CNS selectivity II. In vivo characterization. *J Pharmacol Exp Ther* 1996;**279**:1422–9.
15. Hedberg MM, Svedberg MM, Mustafiz T, et al. Transgenic mice overexpressing human acetylcholinesterase and the Swedish amyloid precursor protein mutation: effect of nicotine treatment. *Neuroscience* 2008;**152**:223–33.
16. Unger C, Svedberg MM, Yu WF, Hedberg MM, Nordberg A. Effect of subchronic treatment of memantine, galantamine, and nicotine in the brain of Tg2576 (APPswe) transgenic mice. *J Pharmacol Exp Ther* 2006;**317**:30–6.

hippocampal neurons, soluble Aβ can bind to α7 nAChRs and block nAChR-mediated signals.[106,107] Dual effects were reported in two-electrode voltage clamp recordings on *Xenopus laevis* oocytes expressing rat α7 nAChRs (wild-type and mutant), characterizing the response to Aβ peptide 1–42 applied at concentrations ranging from 1 pM to 100 nM. α7 nAChRs were activated at a picomolar concentration range, whereas higher concentrations resulted in possible receptor desensitization and antagonistic effects.[108] Similar results (agonistic or antagonistic of nAChR, depending on the concentration) were observed *in vitro* in primary cultures of rat cortical neurons.[109] Lilja et al. evaluated the functional effects of Aβ fibrils and oligomers on α7 nAChRs by measuring intracellular Ca^{2+} levels in PC12 and SH-SY5Y cells. Oligomeric, but not fibrillar, Aβ increased Ca^{2+} levels in neuronal cells, and this effect was attenuated by the α7 nAChR agonist varenicline. Different forms of Aβ can exert neurotoxic effects, which are mediated partly through a blockade of α7 nAChRs, whilst oligomeric Aβ may act as a ligand activating α7 nAChRs, thereby stimulating downstream signaling pathways.[110]

Another pathological hallmark of AD is accumulation of phosphorylated tau protein that aggregates intracellularly in NFTs. Interaction between nAChRs and tau and its impact on AD pathology and cholinergic systems dysfunction is less studied compared with interactions between nAChRs and Aβ. Levels of phosphorylated tau increased in SH-SY5Y human neuroblastoma cells expressing α3, α5, α7, β2, and β4 nAChR subunits after treatment with 10^{-5} M nicotine and 10^{-7} M epibatidine (nAChR agonist) for 3 days; this increase was paralleled and possibly mediated by upregulation of nAChRs.[111] In SK-N-MC neuroblastoma cells and in hippocampal synaptosomes, Aβ-induced tau phosphorylation was attenuated after blocking α7 nAChRs either by antagonist or by antisense RNA.[112] In a transgenic mouse model of AD recapitulating both tau and Aβ pathologies, chronic nicotine administration in drinking water increased the expression of α7 nAChRs and the amount of phosphorylated tau via p38 MAP kinase.[113] A recent article by Yin et al. investigated the interaction between the α4 subunit and overexpressed full-length wild-type human tau in primary hippocampus embryonal rat neurons. Decrease in amount and function of α4 subunit was detected by Western blot analysis and electrophysiology, respectively. Tau-induced degradation of α4 nAChR was attenuated by selective blocking of calpain 2, intracellular Ca^{2+}-dependent cysteine proteases, suggesting a role for this protein in receptor turnover.[114] However, direct interaction between tau and nAChRs has not been reported. Additional work is needed to determine the exact relationship between phosphorylation and levels of tau protein and nAChRs.

2.3.5 Summary

Smoking is a recognized risk factor for development of AD. However, several epidemiological studies from the early 1990s suggested that there was an inverse relationship between smoking and incidence of AD. Many meta-analyses of these studies have been performed in more recent years, and there would appear to be a relatively consistent finding that smoking is associated with a small increase in risk of AD.[17] Because of cholinergic neuronal loss in AD patients, ACh release in the brain is dramatically reduced. Nicotine treatment upregulates nAChRs, leading to the hypothesis that nicotine may improve the condition of some AD patients. Preclinical data showed both positive and negative results following administration of nicotine in different animal models of AD. Several clinical trials have been conducted, with some positive results such as sustained improvement in attention and verbal learning. Criticism of the study designs and the number of subjects enrolled has been raised, and nicotine could not be recommended as a treatment for AD. Moreover, chronic nicotine exposure can lead to loss of nicotinic functional activity as a result of the persistent deactivation of nAChR receptors (i.e., nAChRs desensitization), a mechanism that might explain, in part, the modest or absent effect observed in AD patients. However, interesting positive results from nicotine administration via transdermal patch were found in clinical trials with MCI patients, a state considered transitional toward AD. More clinical studies with larger numbers of subjects are needed for more sound conclusions about the therapeutic potential of nicotine in MCI and AD. Nicotine may be investigated in conjunction with other nonnicotine compounds from tobacco or other medicinal plants for stronger effects in the context of AD pathology.

Chapter 3

Multiple Sclerosis

3.1 SYMPTOMS AND EPIDEMIOLOGY

Multiple sclerosis (MS) is a chronic inflammatory demyelinating disease of the central nervous system (CNS) that affects approximately 0.1%–0.2% of the population in North America and Europe. Women are affected two to three times more often than men. Considerable evidence indicates that autoimmunity plays an important role in the etiology of MS,[115] whereby the patient's immune system attacks and damages myelin, oligodendrocytes (myelin-producing cells), and the underlying nerve fiber. T-cells are activated and infiltrate the CNS through blood vessels. This results in a state of chronic inflammation and damaged nerve fibers (axons) and leads to the recruitment of more immune cells. Clinical presentation of MS varies among patients and depends on the location of the affected nerve fibers. The most common symptoms are sensory loss, motor spinal cord symptoms (numbness or weakness in one or more limbs), autonomic spinal cord symptoms (bladder, bowel, and sexual dysfunction), cerebellar symptoms, optic neuritis or blurred vision, subjective cognitive difficulties and depression.[116]

Based on its histopathology and pathogenic mechanisms, four different clinical courses of MS have been identified. The most common is relapsing-remitting MS (RRMS), in which clearly defined attacks (relapses) of worsening neurological function occur from time to time and are followed by partial or complete remission. Neurological disabilities in the majority of these patients accumulate without proper recovery, leading to transition to secondary-progressive MS (SPMS), with severe nerve damage. In other patients, the progressive course starts at the beginning of the clinical manifestation of MS, with no distinct relapses or remissions. This is known as primary-progressive MS (PPMS) and has a tendency toward more spinal cord lesions. There is also a progressive form termed progressive-relapsing MS (PRMS), which encompasses occasional relapses.[117,118]

3.2 MOLECULAR MECHANISMS

The etiology of MS is currently unknown, and its pathogenesis is only partially understood. Complex genetics and environmental factors determine individual susceptibility to disease development. Twenty-nine disease-related genes, mainly linked to immune system responses, were identified by the International Multiple Sclerosis Genetics Consortium and the Wellcome Trust Case Control

Nicotine and Other Tobacco Compounds in Neurodegenerative and Psychiatric Diseases.

https://doi.org/10.1016/B978-0-12-812922-7.00003-2

Consortium 2 in 2011.[119] The most prominent environmental factors are early-life Epstein-Barr virus infection, ultraviolet light exposure, and vitamin D status.[120]

Published results indicate that immune mechanisms play an essential role in driving the disease process. The pathological hallmarks of MS are neural inflammation, demyelination, remyelination, neurodegeneration, and glial scar formation, which occur either focally or diffusely throughout the white and gray matter of the brain and spinal cord. These pathological features are present in all clinical courses of MS.[121] In the early (acute) stage of MS, lymphocytes (predominantly T-cells) are activated in the peripheral circulation by processed peptides that mimic some CNS antigens (e.g., myelin-associated glycoprotein, myelin oligodendroglia glycoprotein, and proteolipid protein). Subsequently, the activated T-cells migrate to the brain parenchyma and become reactivated by antigen-presenting cells (astrocytes or microglial cells). Numerous cytokines, chemokines, nitric oxide (NO), reactive oxygen species (ROS), glutamate, and free radicals induce and regulate multiple critical disease-associated cellular functions and are produced through further stimulation of myelin-reactive T-cells. This positive feedback loop acts on effector cells and results in myelin, oligodendrocyte, and neuronal damage.[122] In response to this damage, oligodendrocytes perform incomplete and inappropriate remyelination.[123] An additional damage response is yet another complex biological process: glial scar formation by gliosis. Gliosis involves proliferation or hypertrophy of several different types of glial cells, including astrocytes, microglia, and oligodendrocytes.[124] To add to the etiologic complexity of MS, a growing body of scientific evidence suggests that inflammation-driven pathological processes are accompanied by mitochondrial injury/dysfunction-driven neurodegeneration.[121,125,126]

3.3 IMPACT OF SMOKING, SNUS, AND NICOTINE

3.3.1 Smoking and MS: Epidemiological Evidence

There is increasing evidence that tobacco smoking is an associative factor in MS (Table 1). Numerous studies have investigated the effect of smoking before the onset of MS and its impact on disease course.[14] Furthermore, a dose-response association between smoking and MS was identified from two Swedish population-based case-control studies of 7883 cases and 9264 controls. Specifically, MS risk was compared in subjects with different smoking habits by calculating ORs and 95% CIs. A clear dose-response association between cumulative smoking dose and MS risk was observed (P value for trend $<10^{-35}$). Both smoking duration and intensity independently contributed to this increased risk of MS. Nonetheless, the effect of smoking declined a decade after smoking cessation, regardless of the cumulative dose of smoking. Age at smoking initiation did not affect the association between smoking and MS in this study, and smoking increased the risk of MS in a dose-responsive manner. However, contrary to several other MS risk factors that appear to affect risk only if the exposure takes place during a specific period in life, smoking affected MS risk regardless of age at exposure.[127] Genetic

TABLE 1 Epidemiological Studies on Smoking and Multiple Sclerosis—Meta-Analyses

Studies Included	Factors	OR[a]/RR[b] (95% CI[c])	References
24 Case-control and 5 cohort studies between 1980 and 2015	Ever smoker	Conservative model, 1.55 (1.48–1.62) Nonconservative model, 1.57 (1.50–1.64)	1
20 Case-control and 6 cohort studies between 1965 and 2014	Ever smoker	1.51 (1.38–1.65)	2
11 Case-control and 3 cohort studies between 1960 and 2010	Ever smoker	Conservative model, 1.48 (1.35–1.63) Nonconservative model, 1.52 (1.39–1.66)	3
5 Case-control and 1 cohort studies between 1964 and 2006	Ever smoker	1.34 (1.17–1.54)	4

[a] *OR, odds ratio.*
[b] *RR, relative risk.*
[c] *CI, confidence interval.*
1. Zhang P, Wang R, Li Z, et al. The risk of smoking on multiple sclerosis: a meta-analysis based on 20,626 cases from case-control and cohort studies. *PeerJ* 2016;**4**:e1797.
2. O'Gorman C, Broadley S. Smoking and multiple sclerosis: evidence for latitudinal and temporal variation. *J Neurol* 2014;**261**:1677–83.
3. Handel AE, Williamson AJ, Disanto G, Dobson R, Giovannoni G, Ramagopalan SV. Smoking and multiple sclerosis: an updated meta-analysis. *PLoS One* 2011;**6**:e16149.
4. Hawkes CH. Smoking is a risk factor for multiple sclerosis: a metanalysis. *Mult Scler* 2007;**13**:610–5.

factors may also influence susceptibility to MS development in smokers. The *N*-acetyltransferase 1 gene (*NAT1*) is involved in the metabolism of aromatic and heterocyclic aromatic amines (including many tobacco smoke constituents), with a variant associated with increased risk of acquiring MS.[128]

Table 1 summarizes meta-analysis on smoking and MS. A case-control study by Hedstrom et al. (not included in the last meta-analysis presented) also reported that smoking contributes to MS risk in a dose-dependent manner (odd ratio (OR), 1.6; 95% confidence interval (CI) 1.2–2.0 for <10 pack years smokers; OR, 2.0; 95% CI 1.6–2.8 for ≥10 years smokers).[129]

3.3.2 Snus and MS: Epidemiological Evidence

Findings on the impact of smoking and snus consumption on MS risk development in Swedish populations suggest that the use of Swedish snus is not associated with increased MS risk; this indicates that nicotine may not be the substance responsible for increased MS risk development among smokers.[15] Hedström and colleagues further investigated this correlation and reported an unexpected finding, specifically, that snus-takers show a decreased risk of developing MS compared with those who have never used moist snus (OR, 0.83; 95% CI, 0.75–0.92). Additionally, subjects who combined smoking and snus use had a significantly lower risk for MS than smokers who had never used moist snus, even after adjusting for the amount of smoking.

3.3.3 Nicotine and MS: Preclinical Evidence

Several studies have suggested that nicotine does not promote more severe symptoms of experimental autoimmune encephalomyelitis (EAE), but rather inhibits disease development.[130,131] Cigarette smoke constituents other than nicotine appear to be responsible for the clear positive correlation between smoking and MS development and progression. Table 2 summarizes studies on the

TABLE 2 Nicotine in Multiple Sclerosis—Preclinical Studies

Model	Nicotine Treatment	Effect	References
MOG[a] peptide-induced EAE[b] mice	Nicotine osmotic mini-pump (0.39 mg/day) for 28 days	Delayed onset of EAE and reduced disease severity	1
MOG peptide-induced EAE mice	Nicotine osmotic mini-pump for 14 or 28 days	Reduced severity of EAE Prevention of disease exacerbation if administered after symptom development	2
	Nonnicotinic cigarette smoke condensate (osmotic mini-pump)	Accelerated and increased adverse clinical symptoms during the early stage of EAE	
MOG peptide-induced EAE mice	Osmotic mini-pump implanted subcutaneously on the right side of the back (12 μL nicotine/day, equivalent to 0.39 mg nicotine-free base per mouse per day)	Significantly delayed and attenuated inflammatory and autoimmune responses to myelin antigens	3

[a] *MOG, myelin oligodendrocyte glycoprotein.*
[b] *EAE, experimental autoimmune encephalomyelitis.*

1. Hao J, Simard AR, Turner GH, et al. Attenuation of CNS inflammatory responses by nicotine involves alpha7 and non-alpha7 nicotinic receptors. *Exp Neurol* 2011;**227**:110–9.
2. Gao Z, Nissen JC, Ji K, Tsirka SE. The experimental autoimmune encephalomyelitis disease course is modulated by nicotine and other cigarette smoke components. *PLoS One* 2014;**9**:e107979.
3. Shi F-D, Piao W-H, Kuo Y-P, Campagnolo DI, Vollmer TL, Lukas RJ. Nicotinic attenuation of central nervous system inflammation and autoimmunity. *J Immunol* 2009;**182**:1730–9.

effects of nicotine in an *in vivo* EAE model. Delayed onset of EAE and reduced disease severity were observed when nicotine was administered by osmotic pump. For example, Gao and colleagues reported that nicotine reduced demyelination, increased body weight, and attenuated microglial activation in a mouse EAE model. Nicotine administration after the development of EAE symptoms prevented further disease exacerbation. In contrast, the remaining components of cigarette smoke, delivered as cigarette smoke condensate, accelerated and increased adverse clinical symptoms during the early stages of EAE.[131]

3.3.4 nAChRs in MS

The role of nicotinic receptors (for details about nicotinic receptors see Section 9.1.2) in MS is incompletely understood and is informed mostly from animal studies and from various *in vitro* models. An impact of nAChRs during immune cell ontogeny has been observed. In fetal thymus organ culture, exposure to low concentrations of nicotine (10^{-18} M) over 12 days modulated development of immature T-cells.[132] The presence of different nAChRs during the course of B-cell maturation in C57BL/6 mice appears to depend on the location where B-cells originate. The highest expression of α4α5/β2 receptors was observed in immature newly generated B-lymphocytes of the bone marrow, while the number of α7α5/β4 subunit-containing receptors increased with B-cell maturation in the spleen. The role of α7-containing receptors appears critical for the propagation of mature B-lymphocytes in the spleen.[133] Nicotine exposure suppresses the T-cell response and alters the differentiation, phenotype, and functions of antigen-presenting cells, including mouse bone marrow-derived precursor dendritic cells[134,135] and macrophages.[136]

Nicotine exposure has been shown to protect against inflammation and immune attack. The expression of nAChRs by neurons, microglia, and astrocytes and in the periphery (blood cells including lymphocytes, macrophages, mast cells, dendritic cells, and basophils) suggests possibly diverse mechanisms by which natural ACh signaling could modulate immune responses.[137] Nicotinic AChR signaling is thought to have beneficial, immunosuppressive consequences via what is termed the cholinergic antiinflammatory pathway, the cytokine-suppressing mechanism of the inflammatory responses through the vagus nerve neuronal circuits, which prevents immune-mediated damage.[130,138] In primary human macrophages, ACh inhibits the release of tumor necrosis factor (TNF) and other cytokines through a posttranscriptional mechanism that depends on α-bungarotoxin-sensitive nAChRs.[5] The nature of receptors responsible for those effects was identified later, and the role of α7 nAChR was characterized[139] and confirmed.[140]

The role of individual nAChR subunits was evaluated using an array of knockout (KO) mice for different nAChR subunits. Early studies suggested that the cholinergic antiinflammatory pathway acts via α7 nAChRs.[139,141] Other nAChR subtypes are involved as well, since nicotine retains the ability to modulate multiple immunologic functions by reducing lymphocyte infiltration into the CNS, inhibiting autoreactive T-cell proliferation and helper T-cell cytokine production, downregulating costimulatory protein expression in myeloid cells, and increasing the differentiation and recruitment of regulatory T-cells in nAChR $\alpha7$ subunit KO mice.[142] Using $\alpha9$ and $\beta2$ KO animals for EAE, Simard and colleagues[143] observed that signaling via $\beta2$-containing nAChRs is important for recovery after the disease symptoms develop, rather than for nicotine-induced protection, since recovery is prevented in $\beta2$ KO mice, while nicotine retains the ability to diminish the severity of EAE symptoms. Alpha-9 subunit gene deletion alone delays the onset and reduces the severity of EAE. These findings suggest that $\alpha9$-containing nAChRs are naturally involved in endogenous pro-inflammatory mechanisms required for disease initiation and evolution, likely via the endogenous agonist, ACh. Nicotine administration had no effect on progression of EAE in $\alpha9$ KO mice, which is not surprising since nicotine is antagonistic of nAChRs containing the $\alpha9$ subunit.[144] Further experiments with $\alpha7$ and $\alpha9$ KO mice confirmed previous findings that both nAChRs are important for the modulation of immune responses.[145] Obviously, different nAChR subtypes play disease-protecting or exacerbating roles in various stages of disease progression and recovery.

3.3.5 Summary

Smoking has been recognized as a risk factor in the development of MS. Interestingly, snus consumers and combined snus and cigarette consumers appear to have a reduced risk of developing MS compared with smokers who have never used snus, based on several studies in the Swedish population. From preclinical studies using the EAE model, nicotine appears to delay disease onset and reduce severity. Further epidemiological studies on noncombustible tobacco and nicotine-containing products will certainly give new interesting insights into correlations between MS and consumption of those products. Given the importance of various nAChRs in modulating disease exacerbation or ameliorating inflammatory and/or immune processes, nAChRs should be further investigated as possible drug targets.

Chapter 4

Schizophrenia

4.1 SYMPTOMS AND EPIDEMIOLOGY

Schizophrenia is a severe brain disorder in which people interpret reality abnormally. Patients present with a combination of hallucinations, delusions, and extremely disordered thinking and behavior.[146] Schizophrenia is a debilitating psychiatric disorder characterized by symptoms described in the literature as positive (e.g., hallucinations and delusions), negative (e.g., depressed mood and social withdrawal), and cognitive (e.g., deficits in information processing, attention, working memory, and executive function). Symptoms typically develop gradually, beginning in young adulthood. Schizophrenia is a chronic condition, requiring lifelong treatment. Approximately 0.3%–0.7% of people are affected by schizophrenia during their lifetime.[147]

4.2 MOLECULAR MECHANISMS

The development of schizophrenia is multifactorial, and both genes and environmental impact play a role. Current data suggest neurodevelopmental or environmental causes such as viral infections and prenatal/perinatal complications. Predisposition to schizophrenia is affected by early prenatal or perinatal insults that lead to defects in early brain development. Schizophrenic brains lack the "normal" leftward anterior cingulate cortex (ACC) sulcal asymmetry, as a result of reduced folding in the left ACC. Prenatal exposure to viral infections such as influenza and poliovirus, poor prenatal nutrition, adverse obstetric events, and cannabis smoking during adolescence are examples of environmental factors that may increase the risk of schizophrenia.[148]

Schizophrenia manifestations are more common in some families, and the disease is likely a complex disorder where multiple genes are involved. Risk genes may impact neurodevelopment and neuronal functions, increasing the risk for schizophrenia. A number of genes have been identified as risk genes through various genetic analyses, including dystrobrevin binding protein 1 (*DTNBP1*), proline dehydrogenase (*PRODH*), D-amino acid oxidase activator (*DAOA*), trace amine-associated receptor 6 (*TAAR6*), zinc-finger DHHC type-containing 8 (*ZDHHC8*), neuregulin 1 (*NRG1*), catechol-O-methyltransferase (*COMT*), disrupted in schizophrenia 1 (*DISC1*), and neuropeptide Y (NPY).[149] However, the exact function of some of these genes is not known. Mutations

Nicotine and Other Tobacco Compounds in Neurodegenerative and Psychiatric Diseases.

https://doi.org/10.1016/B978-0-12-812922-7.00004-4

in the α7 nAChR gene and reduction in α7 nAChR density in the hippocampus of schizophrenics have also been linked to schizophrenia pathophysiology.[150] Alpha-7 receptors appear to be of particular importance in the pathology of schizophrenia; postmortem receptor binding studies on schizophrenia patients showed a reduction in α7 receptors in hippocampi and thalamic nuclei.[151] Furthermore, α7 receptor mRNA and protein levels are significantly lower in schizophrenic nonsmokers compared with control nonsmokers and are brought to control levels in schizophrenic smokers.[152] Details on the implication of various nAChRs in schizophrenia are available in a review by Featherstone and Siegel.[153]

Other than altered nAChR levels in schizophrenia patients, which impact the cholinergic neurotransmitting system, impairments in *N*-methyl-D-aspartate (NMDA), dopamine D2, and GABA neurotransmission have also been described.[148,154] The original dopamine hypothesis states that hyperactive dopamine transmission results in schizophrenic symptoms. Dopamine receptor blockade by chlorpromazine and haloperidol, proposed in the early 1960s, has been a cornerstone of psychiatric treatment.[155] The revised dopamine hypothesis takes into consideration positive and negative symptoms. Positive symptoms result from increased subcortical release of dopamine, which augments D2 receptor activation, and negative symptoms reduce D1 receptor activation in the prefrontal cortex. This hypothesis proposes hyperactive dopamine transmission in the mesolimbic areas and hypoactive dopamine transmission in the prefrontal cortex in schizophrenia patients. Furthermore, dopaminergic and serotonergic deviations are known to contribute significantly to both the positive and negative symptoms of schizophrenia. Dopamine dysregulation is also observed in brain regions such as the amygdala and prefrontal cortex, which are important for emotional processing.[156]

4.3 IMPACT OF SMOKING AND NICOTINE

4.3.1 Smoking and Schizophrenia: Epidemiological Evidence

Smoking is strongly associated with schizophrenia; schizophrenics are much more likely to smoke than those without the disease. A meta-analysis by de Leon and Diaz estimated a 65%–85% prevalence of tobacco smoking in schizophrenia patients.[157] Tobacco products are heavily used by schizophrenics, which is believed to be an attempt to improve cognition, such as information (sensory) processing and attention deficits.[158] Unlike relationships between smoking and above described studies where the risk of development of diseases in smokers is presented, Table 1 summarizes data on the likelihood that schizophrenia patients would smoke. Another case-control study that was published in 2013 (not included in the presented meta-analysis) is reporting similar results, OR 1.83 and 2.52 for lifetime and current male patients, respectively ($P < 0.001$).[159]

TABLE 1 Epidemiological Studies on Smoking and Schizophrenia—Meta-Analyses

Studies Included	Factors	OR[a]/RR[b] (95% CI[c])	References
10 Studies before 2001	Ever smoker	6.04 (3.03–12.02)	1
42 Studies between 1995 and 2004	Current smoker	5.9 (4.9–5.7)[d]	2

[a] *OR, odds ratio.*
[b] *RR*, relative risk.
[c] *CI*, confidence interval.
[d] In the cited article, the ORs were differently stated in the abstract, OR=5.9 (4.9 – 5.7), and in the main text, OR=5.3 (4.9 – 5.7). The former one was adapted here.
1. Myles N, Newall HD, Curtis J, Nielssen O, Shiers D, Large M. Tobacco use before, at, and after first-episode psychosis: a systematic meta-analysis [CME]. *J Clin Psychiatry* 2012;**73**(1):478–5.
2. de Leon J, Diaz FJ. A meta-analysis of worldwide studies demonstrates an association between schizophrenia and tobacco smoking behaviors. *Schizophr Res* 2005;**76**:135–57.

4.3.2 Nicotine and Schizophrenia: Clinical Evidence

In schizophrenia patients, nicotine decreases some severe psychiatric symptoms and improves sensory gating and prepulse inhibition (PPI).[160] PPI is a neuropsychological phenomenon in which the motor response to a startling stimulus (pulse) is significantly reduced when that stimulus is closely preceded in time by another usually weaker stimulus (prepulse). Nicotine increases dopamine release; therefore, it is hypothesized that nicotine delivered through smoking helps correct dopamine deficiency in the prefrontal cortex and relieve negative symptoms.[161,162] A number of clinical studies have been conducted and are currently ongoing, in which nicotine as patch, gum, or a nasal spray was used by schizophrenia patients (Table 2). Based on the data reported so far, nicotine appears to improve some aspects of cognition, but the absence of effects is reported as well. Mixed results were reported by Harris and colleagues where 20 schizophrenics, 10 smokers and 10 nonsmokers, were assessed following the administration of nicotine gum and placebo gum. The Repeatable Battery for the Assessment of Neuropsychological Status test was administered. Nicotine affected only the Attention Index, and there were no effects on learning and memory, language, or visuospatial/constructional abilities. Attentional function was increased in nonsmokers but decreased in nicotine-abstinent smokers after nicotine administration.[163]

4.3.3 Nicotine and Schizophrenia: Preclinical Evidence

The effects of nicotine on schizophrenia have also been investigated in a preclinical setting. Sensorimotor gating refers to the modulation of motor responses observed when multiple sensory stimuli are presented in rapid succession.[164] Prepulse inhibition (PPI) inhibition in rats or mice is a commonly used approach in drug development. Findings on the impact of nicotine on PPI *in vivo* are mixed, showing improvements or no effects (summarized in Table 3).

TABLE 2 Clinical Trials on Nicotine and Schizophrenia

Study Design	Sample Description	Nicotine Treatment	Outcome	References
Placebo-controlled crossover	31 Patients (current smokers)	Nicotine nasal spray (10 mg/kg) or placebo spray administered over 2–2.5 h	Nicotine nasal spray modestly improved selected aspects of cognition	1
Placebo-controlled double-blind crossover	30 patients (current smokers)	Transdermal nicotine patch (21 mg) or placebo patch, separated by at least 24 h	Trend toward reduced bradykinesia (rigidity induced by haloperidol treatment)	2
Placebo-controlled	20 Patients (10 smokers and 10 nonsmokers)	Subjects administered either nicotine gum (2 mg, ×3) or placebo gum to chew for 10 min	Modest effects on attention—decline in smokers and increase in nonsmokers No effect in other cognitive functions	3
Placebo-controlled double-blind crossover	27 Male patients (current smokers)	Two sessions of nicotine nasal spray and two sessions of placebo spray	Small effect of cognitive enhancement on attention and visual-spatial memory	4
Randomized, double-blind, placebo-controlled	28 Patients and 32 healthy controls (all nonsmokers)	Subjects administered two nicotine patches (7 mg) and placebo patches Active and placebo doses separated by 2 weeks	Improved attentional performance in both groups, with greater improvements in inhibition of impulsive responses in patients	5
Randomized, double-blind, placebo-controlled	Patients and normal controls	Nicotine patch (7 mg for nonsmokers and 14 mg for smokers) or placebo	Significant effect on error percentage in antisaccade task	NCT01315002[a]

[a]*NCT, National Clinical Trial.*

1. Smith RC, Singh A, Infante M, Khandat A, Kloos A. Effects of cigarette smoking and nicotine nasal spray on psychiatric symptoms and cognition in schizophrenia. *Neuropsychopharmacology* 2002;**27**:479–97.
2. Yang YK, Nelson L, Kamaraju L, Wilson W, McEvoy JP. Nicotine decreases bradykinesia-rigidity in haloperidol-treated patients with schizophrenia. *Neuropsychopharmacology* 2002;**27**:684–6.
3. Harris JG, Kongs S, Allensworth D, et al. Effects of nicotine on cognitive deficits in schizophrenia. *Neuropsychopharmacology* 2004;**29**:1378–85.
4. Smith RC, Warner-Cohen J, Matute M, et al. Effects of nicotine nasal spray on cognitive function in schizophrenia. *Neuropsychopharmacology* 2006;**31**:637–43.
5. Barr RS, Culhane MA, Jubelt LE, et al. The effects of transdermal nicotine on cognition in nonsmokers with schizophrenia and nonpsychiatric controls. *Neuropsychopharmacology* 2008;**33**:480–90.

TABLE 3 Nicotine and Schizophrenia—Preclinical Studies

Model of PPI[a] Deficits	Nicotine Treatment	Effect	References
Amphetamine-induced in Sprague-Dawley rats	Nicotine infusion to LHb[b] (25 and 50 mg)	Dose-dependent attenuation of PPI deficits	1
Amphetamine-induced in male Wistar rats	Nicotine (0.01–0.2 mg/kg, s.c.[c])	Dose-dependent attenuation of PPI deficits	2
Methamphetamine-induced in male ICR mice	Nicotine (0.15–0.5 mg/kg)	Ameliorated PPI deficits, reversed c-fos activation in LGP[d] and PnC[e]	3
PCP[f]-induced in BALB/cByJ mice	Nicotine (0.03–1 mg/kg)	Nicotine (1 mg/kg) selectively reversed PPI deficits in BALB/cByJ mice but not NMRI mice	4
Quinpirole-induced in female Sprague-Dawley rats	Acute nicotine (0.05–0.4 mg/kg)	Acute high dose (0.4 mg/kg) improved PPI deficits	5
	Chronic nicotine (0.2 or 0.4 mg/kg)	Chronic doses (both 0.2 and 0.4 mg/kg) almost normalized deficits	
Dizocilpine (MK801)-induced in female Sprague-Dawley rats	Acute nicotine (0.05–0.4 mg/kg) Chronic nicotine (0.2 or 0.4 mg/kg)	Neither treatment exerted an effect	5

[a] *PPI, prepulse inhibition.*
[b] *LHb, lateral habenula.*
[c] *s.c., subcutaneous.*
[d] *LGP, lateral globus pallidus.*
[e] *PnC, pontine reticular nucleus.*
[f] *PCP, phencyclidine.*

1. Larrauri JA, Burke DA, Hall BJ, Levin ED. Role of nicotinic receptors in the lateral habenula in the attenuation of amphetamine-induced prepulse inhibition deficits of the acoustic startle response in rats. *Psychopharmacology* 2015;**232**: 3009–17.
2. Suemaru K, Yasuda K, Umeda K, et al. Nicotine blocks apomorphine-induced disruption of prepulse inhibition of the acoustic startle in rats: possible involvement of central nicotinic α7 receptors. *Br J Pharmacol* 2004;**142**:843–50.
3. Mizoguchi H, Arai S, Koike H, et al. Therapeutic potential of nicotine for methamphetamine-induced impairment of sensorimotor gating: involvement of pallidotegmental neurons. *Psychopharmacology* 2009;**207**:235–43.
4. Andreasen JT, Andersen KK, Nielsen EØ, Mathiasen L, Mirza NR. Nicotine and clozapine selectively reverse a PCP-induced deficit of PPI in BALB/cByJ but not NMRI mice: comparison with risperidone. *Behav Brain Res* 2006;**167**:118–27.
5. Nespor AA, Tizabi Y. Effects of nicotine on quinpirole and dizocilpine (MK-801)-induced sensorimotor gating impairments in rats. *Psychopharmacology* 2008;**200**:403–11.

For example, Nespor and Tizabi reported different findings after acute and chronic nicotine administration with regard to PPI, depending on how PPI deficits were induced. Nicotine appears not to impact dizocilpine-induced deficits, while quinpirole-induced deficits were partially or fully reverted with both acute and chronic nicotine administration.[165] Mixed results are probably a consequence of differently induced PPI deficits and different doses of nicotine used in different rodent strains.

4.3.4 Summary

A strong association between schizophrenia and smoking was demonstrated in epidemiological studies across 20 countries, whereby people with schizophrenia are much more likely to smoke than those without the disease, probably in an attempt to improve cognition, such as information (sensory) processing and attention deficits. Smoking among schizophrenics is a serious problem and leads to increased rates of mortality, increased risk of cardiovascular disease, and other serious smoking-related diseases, along with reduced treatment effectiveness. Nicotine increases dopamine release; therefore, it is hypothesized that nicotine delivered through smoking helps to correct the dopamine deficiency in the prefrontal cortex and relieve negative symptoms. This was the basis for a number of clinical and preclinical studies investigating the effects of nicotine alone. Both preclinical and clinical data on nicotine in the context of symptom relief yielded mixed results, showing some positive effects or no effects at all. In clinical studies, nicotine was delivered as a nasal spray, patch, or gum. Several clinical trials are ongoing, indicating that the potential of nicotine as a therapeutic agent for schizophrenia is still a topic of interest. More clinical data on the effects of nicotine are needed to support the use of nicotine as a symptom-relief aid for schizophrenia patients. Similarly to AD or PD, nicotine in conjunction with some other medicinal plant constituents (not necessarily from tobacco) may exert more powerful effects in treating schizophrenia.

Chapter 5

Tourette Syndrome

5.1 SYMPTOMS, EPIDEMIOLOGY, AND POTENTIAL CAUSES

Tourette syndrome (TS) is a childhood onset hyperkinetic disorder that manifests in the form of repetitive, sudden, intermittent motor, and/or phonic tics.[166,167] Tics linked to TS are largely involuntary and can range from simple tics, which are simple repetitive actions involving limited muscle groups, to complex tics, which are distinct and coordinated movements involving several muscle groups. The tics generally occur in bouts, many times per day, every day. Many individuals are able to voluntarily suppress their tics; however, this can cause significant discomfort and stress, and the urge eventually becomes uncontrollable after a period of suppression.[168]

As TS appears to be a disease of complex and heterogeneous origins, there are numerous treatment regimes, each with efficacy limited to specific patient populations. These include the following:

- *Dopamine antagonists*. Drugs such as Haldol (haloperidol) and Orap (pimozide) have been shown to control tics.
- *Botulinum injections*. Simple motor tics may be controlled by Botox injection to the affected muscle.
- *Other treatment regimes*. Medications to control comorbidities such as depression and muscle control may be treated pharmacologically and offer some improvement through stress relief.
- *Adjunct therapies*. These include behavioral therapy to help patients recognize premonitory urges (sensation preceding tic release).

TS affects approximately 1% of school-aged children, with approximately 37% of those affected facing moderate to severe forms. There appears to be a gender bias, whereby the ratio in males to females is approximately 5:1, although this resolves into an approximately 1:1 ratio in adulthood. The first tics generally appear at a young age, between 2 and 15 years old; the average age of onset is 6 years. In many cases, tics lessen or become manageable after the teen years, with a period of maximal tic severity occurring between 11 and 14 years. A minority (approximately 20%–30%) of TS patients will continue to present tics in adulthood, and there is currently no predictive method to distinguish this population from the patients who ultimately gain control over their tics.[168]

Nicotine and Other Tobacco Compounds in Neurodegenerative and Psychiatric Diseases.

https://doi.org/10.1016/B978-0-12-812922-7.00005-6

The pathophysiology of TS appears to involve a complex dysregulation of the dopaminergic neurotransmitter system in basal ganglia-related circuits, especially hyperactive dopaminergic innervation.[169] Evidence from postmortem studies of TS patients have shown increased numbers of striatal[170] and cortical[171] dopamine 2 receptors and altered dopamine metabolism in the basal ganglia.[172,173]

Substantial evidence suggests that both environmental and genetic factors contribute to the development and clinical expression of TS. The precise inheritance pattern is as yet unclear. Slit and NTRK-like family member 1 (*SLITRK1*) is one of several genes thought to be associated with TS.[174] However, molecular genetic testing of *SLITRK1* is of little or no clinical relevance based on current knowledge.[175] Twin studies strongly support a significant role for nongenetic contributions. Nonetheless, identifying specific environmental causes or contributors has proved as difficult as identifying specific disease-related genes.

5.2 IMPACT OF SMOKING AND NICOTINE

5.2.1 Smoking and TS: Epidemiological Evidence

One of the most widely investigated nongenetic contributors to TS development is the role of maternal smoking (*in utero* exposure). In a Finnish study of 767 children, prenatal maternal smoking was associated with TS syndrome when comorbid with attention deficiency hyperactivity disorder (ADHD), but there was no association between maternal smoking during pregnancy and TS syndrome without comorbid ADHD.[176] In an analysis of 73,073 pregnancies in the Danish National Birth Cohort, heavy smoking was associated with a 66% increased risk for TS and chronic tic disorder. Heavy smoking (more than 10 cigarettes per day) was associated with a twofold increased risk for TS and chronic tic disorder with comorbid ADHD, and both light and heavy smoking were associated with a more than twofold increased risk for TS and chronic tic disorder with any non-ADHD psychiatric comorbidity.[177] Reasons for such a complex correlation have not been fully elucidated. Other epidemiological data are summarized in Table 1.

5.2.2 Nicotine and TS: Clinical Evidence

In the context of TS symptoms, nicotine has been tested using patch and gum as administration routes (Table 2). Initially, nicotine gum was shown to significantly decrease tics and improve attention span when used in combination with the dopamine receptor antagonist haloperidol, while nicotine gum alone had little effect.[178,179] Subsequent studies using transdermal nicotine patches suggest a similar reduction in tics in patients taking haloperidol.[180–183] Importantly, an improved outcome following administration via patch over nicotine gum, with longer-lasting relief from TS symptoms was shown.[184] nAChR desensitization has been proposed as a possible explanation for the observed effects in TS patients, and different dynamics of this phenomenon, depending on the route of nicotine administration,

TABLE 1 Epidemiological Studies on Prenatal Smoking and TS[a]

Type of Study	Cases	Controls	Factors	OR[b] (95% CI[c])	References
Case-control	520 (TS alone)	1944	Maternal smoking in first trimester	OR=0.5 (0.2–1.3)	1
			Maternal smoking throughout pregnancy	OR=0.9 (0.7–1.3)	
	203 (TS+ADHD[d])	954	Maternal smoking in first trimester	4.0 (1.2–13.5)	
			Maternal smoking throughout pregnancy	1.7 (1.05–2.7)	
Case-control	40 (TS alone)	65	Maternal smoking	4.6 (0.45–46.6)	2
	60 (TS+ADHD)			8.5 (0.97–75.2)	
Case-control	181 (TS+ADHD)	172 (TS alone)	Maternal smoking	2.43 (1.23–4.82)	3
Cohort	50 (TS alone)	5968	Maternal smoking in first trimester	1.52 (0.75–3.09)	4
	122 (TS+chronic tic disorder)			1.37 (0.86–2.17)	

[a] *TS, Tourette syndrome.*
[b] *OR, odds ratio.*
[c] *CI, confidence interval.*
[d] *ADHD, attention deficiency hyperactivity disorder.*

1. Leivonen S, Chudal R, Joelsson P, et al. Prenatal maternal smoking and Tourette syndrome: a nationwide register study. *Child Psychiatry Hum Dev* 2016;**47**:75–82.
2. Motlagh MG, Katsovich L, Thompson N, et al. Severe psychosocial stress and heavy cigarette smoking during pregnancy: an examination of the pre- and perinatal risk factors associated with ADHD and Tourette syndrome. *Eur Child Adolesc Psychiatry* 2010;**19**:755–64.
3. Pringsheim T, Sandor P, Lang A, Shah P, O'Connor P. Prenatal and perinatal morbidity in children with Tourette syndrome and attention-deficit hyperactivity disorder. *J Dev Behav Pediatr* 2009;**30**:115–21.
4. Mathews CA, Scharf JM, Miller LL, Macdonald-Wallis C, Lawlor DA, Ben-Shlomo Y. Association between pre- and perinatal exposures and Tourette syndrome or chronic tic disorder in the ALSPAC cohort. *Br J Psychiatry* 2014;**204**(1):40–45.

TABLE 2 Clinical Studies on Nicotine and TS[a]

Type of Study	Subject Description	Nicotine Treatment	Outcome	References
Open-label	10 TS patients receiving haloperidol	Nicotine gum	Decrease in tics and improvement of concentration and attention span	1
Open-label	10 TS patients receiving haloperidol	Nicotine gum	Improvement in symptoms	2
Placebo- controlled	19 TS patients, half receiving haloperidol	Nicotine gum	Improvement in haloperidol + nicotine group only	3
Open-label	5 TS patients, not receiving medication	Nicotine patch	Amelioration of symptoms	4
Open-label	20 TS patients	Nicotine patch	Reduction of TS symptoms during and up to 2 weeks after application	5
Double-blind, placebo-controlled	70 TS patients receiving haloperidol	Nicotine patch	Reduction of symptom severity (CGI-I[b] and PGI-I[c] scoring scales; not observed using YGTSS[d] scale)	6

[a] *TS, Tourette syndrome.*
[b] *CGI-I, Clinical Global Impression-Improvement.*
[c] *PGI-I, Patient Global Impression of Improvement.*
[d] *YGTSS, Yale Global Tic Severity Scale.*

1. Sanberg PR, McConville BJ, Fogelson HM, et al. Nicotine potentiates the effects of haloperidol in animals and in patients with Tourette syndrome. *Biomed Pharmacother* 1989;**43**:19–23.
2. McConville BJ, Fogelson MH, Norman AB, et al. Nicotine potentiation of haloperidol in reducing tic frequency in Tourette's disorder. *Am J Psychiatry* 1991;**148**:793–4.
3. McConville BJ, Sanberg PR, Fogelson MH, et al. The effects of nicotine plus haloperidol compared to nicotine only and placebo nicotine only in reducing tic severity and frequency in Tourette's disorder. *Biol Psychiatry* 1992;**31**:832–40.
4. Dursun SM, Reveley MA. Differential effects of transdermal nicotine on microstructured analyses of tics in Tourette's syndrome: an open study. *Psychol Med* 1997;**27**:483–7.
5. Silver AA, Shytle RD, Philipp MK, Sanberg PR. Case study: long-term potentiation of neuroleptics with transdermal nicotine in Tourette's syndrome. *J Am Acad Child Adolesc Psychiatry* 1996;**35**:1631–6.
6. Silver AA, Shytle RD, Philipp MK, Wilkinson BJ, McConville B, Sanberg PR. Transdermal nicotine and haloperidol in Tourette's disorder: a double-blind placebo-controlled study. *J Clin Psychiatry* 2001;**62**:707–14.

may be responsible for the discrepancies in results obtained with gum versus patch. An acute exposure to nicotine from chewing nicotine gum desensitized nAChRs only for a few hours, while a continuous 24h exposure to nicotine through a patch may desensitize nAChRs for a considerable length of time, even following patch removal. Due to the short half-life of nicotine, alternative compounds have been investigated, which lead to renewed interest in a decades-old blood pressure medicine, mecamylamine, which is also known to act as a nAChR antagonist. Mecamylamine was well tolerated by the participants of a clinical trial and showed some reduction in motor and vocal tics and an improvement in mood and behavioral disturbances but was not effective as a monotherapy.[184]

5.2.3 Nicotine and TS: Preclinical Evidence

A limited number of animal studies (Table 3) have been conducted to evaluate nicotine effects in TS models. The impact of nicotine on head twitching

TABLE 3 Nicotine and TS[a]—Preclinical Studies

Model	Nicotine Treatment	Effect	References
DOI[b]-induced head twitching in male Albino mice	Acute study—nicotine salt (0.5–5 mg/kg, intraperitoneal), pretreatment	Dose-dependent reduction in head twitching	1
	Chronic study—nicotine salt (1.5 mg/kg, intraperitoneal), once daily for 10 days	Reduced head-twitching frequency	
DOI-induced head twitching in male MF1 mice	Acute study—nicotine (0.4, 0.8, or 1.6 mg/kg subcutaneous) pretreatment	Dose-dependent reduction in head twitching, with significance at 1.6 mg/kg Effect blocked by mecamylamine	2
	Chronic study—nicotine (1.6 mg/kg subcutaneous, b.i.d.[c]) for 7 days	Significant increase in head-twitching frequency	

[a]*TS, Tourette syndrome.*
[b]*DOI, (1-)2,5-dimethoxy-4-iodophenyl-2-aminopropane.*
[c]*b.i.d., bis in die, twice per day.*

1. Tizabi Y, Russell LT, Johnson M, Darmani NA. Nicotine attenuates DOI-induced head-twitch response in mice: implications for Tourette syndrome. *Prog Neuro-Psychopharmacol Biol Psychiatry* 2001;**25**:1445–57.
2. Gaynor CM, Handley SL. Effects of nicotine on head-shakes and tryptophan metabolites. *Psychopharmacology* 2001;**153**:327–33.

induced by injection of the 5-HT2A/2C receptor agonist (1-)2,5-dimethoxy-4-iodophenyl-2-aminopropane (DOI) in male albino ICR mice was evaluated by Tizabi et al. The authors assessed the effects of acute and chronic pretreatment of nicotine on head-twitching frequency and reported a reduction in this parameter in both administration regimes. Gaynor et al. reported different effects of acutely and chronically administered nicotine on head-twitching frequency. Acute administration of nicotine or the nicotinic agonist epibatidine dose-dependently attenuated head twitching induced by DOI. This attenuation was inhibited by the nicotinic receptor antagonist mecamylamine. Fifteen hours after the second of two daily injections of nicotine (1.6 mg/kg for 7 days), the frequency of spontaneous and DOI-induced head twitching was significantly potentiated.

5.2.4 Summary

Prenatal maternal smoking has been reported to correlate with TS, especially when comorbid with ADHD. Reasons for such a complex correlation have not been fully elucidated. Nicotine administered by gum was shown to significantly decrease tics and improve attention span when used in combination with the neuroleptic haloperidol, while nicotine gum alone had little effect. Nicotine patch was found to have persistent therapeutic benefits sometimes lasting days after patch removal. The pathophysiology of TS is thought to involve a dysfunction of basal ganglia-related circuits and, in particular, hyperactive dopaminergic innervation. nAChR desensitization possibly leading to decrease of dopamine release has been proposed as a possible explanation for the observed effects in TS patients in trials with nicotine gum or a patch. The hypotheses for the effects of nicotine in Parkinson's disease and in TS patients appear to be opposites: the first concerns stimulation of dopamine release in PD patients, while the second concerns desensitization of nAChRs, possibly leading to reduction in dopamine release. This is an example of a wide range of different modalities of nicotine functions via nAChRs that should be investigated further for therapeutic purposes.

Chapter 6

Attention Deficit Hyperactivity Disorder in Children and Adults

6.1 SYMPTOMS, EPIDEMIOLOGY, AND POTENTIAL CAUSES

Attention deficit hyperactivity disorder (ADHD) is a chronic condition that affects millions of children and often continues into adulthood. It is characterized by problems with paying attention, excessive activity, or difficulty controlling behavior, which is not appropriate for a person's age.

Children with ADHD may also struggle with low self-esteem, troubled relationships, and poor performance in school. Symptoms sometimes lessen with age. However, some people never completely outgrow their ADHD symptoms.[146] ADHD affects approximately 5.3% of children and adolescents worldwide. It is estimated that 4.4% of adults in the United States meet the criteria for this disorder.[185,186]

Further, it is estimated that 80% of individuals diagnosed with ADHD have inherited the condition. The polygenetic nature of ADHD indicates that multiple genes jointly contribute to development of this complex disease, and a genetic database for ADHD has recently been created.[187] For the remainder of cases, ADHD may be the residual result of early brain injury/trauma or other impediments to normal brain development. Exposure to chronic low lead levels, prematurity, obstetric complications, malnourishment, and illnesses such as meningitis or encephalitis can all result in learning and attention problems.

6.2 IMPACT OF SMOKING AND NICOTINE

6.2.1 Smoking and ADHD: Epidemiological Evidence

Similar to Tourette syndrome (TS), cigarette smoke exposure *in utero* (summarized in Table 1) reported to correlate with ADHD development. Interestingly, ADHD and TS are often comorbid.[176]

The association between smoking and ADHD is complex. Limited studies with ADHD and smoking have been conducted, because ADHD is often considered a childhood disorder and is thus not included as a psychiatric condition category in adults. Nevertheless, for those studies that have looked at it, ADHD

Nicotine and Other Tobacco Compounds in Neurodegenerative and Psychiatric Diseases.

https://doi.org/10.1016/B978-0-12-812922-7.00006-8

TABLE 1 Epidemiological Studies on Prenatal Smoking and ADHD[a]

Type of Study	Cases	Controls	Factors	OR[b]/RR[c] (95% CI[d])	References
Cross-sectional	4463	NA[e]	Maternal smoking	1.44 (1.06–1.96)	1
			Paternal smoking	1.17 (0.92–1.49)	
Cross-sectional	32	NA	Maternal smoking during pregnancy	3.19 (1.08–9.49)	2
Case-control	12,991	30,071	Maternal smoking during pregnancy	Males, 1.86 (1.53–2.27) Females, 1.67 (1.07–2.61)	3
Case-control	90	270	Maternal smoking during pregnancy	1.8 (0.9–3.6)	4
Case-control	280	242	Maternal smoking during pregnancy	2.1 (1.1–4.1)	5
Case-control	132	139	Maternal smoking during pregnancy	4.4 (1.2–15.5)	6
Case-control	140	120	Maternal smoking during pregnancy	2.7 (1.1–7.0)	7
Cohort	Study population—19,940	NA	Maternal smoking	2.64 (1.45–4.80)	8
			Paternal smoking	1.17 (0.98–1.39)	
Cohort	Study population—84,803	NA	Maternal smoking	1.63 (1.36–1.94)	9
Cohort	Study population—768,227	NA	Maternal smoking during pregnancy		10
			1–9 cig[f]/day	1.62 (1.56–1.69)	
			≥10 cig/day	2.04 (1.95–2.13)	
			Maternal smoking during pregnancy (sibling-matched analysis)		
			1–9 cig/day	0.88 (0.73–1.06)	
			≥10 cig/day	0.84 (0.65–1.06)	

Cohort	Study population—604	NA	Maternal smoking during pregnancy		11
			1–10 cig/day	0.9 (0.4–1.8)	
			>10 cig/day	1.6 (0.8–3.2)	
Cohort	Study population—995	NA	Maternal smoking during pregnancy	2.59 (1.5–4.34)	12
Cohort	Study population—8324	NA	Maternal smoking	1.72 (1.14–2.61)	13
			Paternal smoking	1.43 (0.98–2.07)	
Cohort	Study population—3474	NA	Maternal smoking during pregnancy		14
			Occasionally	1.26 (0.87–1.83)	
			Most days	1.86 (1.31–2.66)	
Cohort	Study population—868,449	NA	Maternal smoking during pregnancy		15
			Sibling-matched analysis	1.2 (0.97–1.49)	
			Unmatched analysis	2.01 (1.90–2.12)	
Cohort	209 Children at risk for ADHD	NA	Maternal smoking	4.0 (1.36–11.12)	16
			Paternal smoking	0.31 (0.06–1.92)	
Cohort	Study population—982,856	NA	Maternal smoking during pregnancy		17
			1–9 cig/day	1.59 (1.49–1.70)	
			≥10 cig/day	1.89 (1.75–2.04)	
			Maternal smoking during pregnancy (sibling-matched analysis)		
			1–9 cig/day	0.96 (0.73–1.33)	
			≥10 cig/day	1.26 (0.95–1.58)	
Cohort	Study population—1342	NA	Maternal smoking during pregnancy	1.53 (1.00–2.35)	18

Continued

TABLE 1 Epidemiological Studies on Prenatal Smoking and ADHD—cont'd

[a] *ADHD, attention deficit hyperactivity disorder.*
[b] *OR, odds ratio.*
[c] *RR, relative risk.*
[d] *CI, confidence interval.*
[e] *NA, not applicable.*
[f] *cig, cigarettes.*

1. Kovess V, Keyes KM, Hamilton A, et al. Maternal smoking and offspring inattention and hyperactivity: results from a cross-national European survey. *Eur Child Adolesc Psychiatry* 2015;**24**:919–29.
2. Koshy G, Delpisheh A, Brabin BJ. Childhood obesity and parental smoking as risk factors for childhood ADHD in Liverpool children. *Atten Defici Hyperact Disord* 2011;**3**:21–8.
3. Silva D, Colvin L, Hagemann E, Bower C. Environmental risk factors by gender associated with attention-deficit/hyperactivity disorder. *Pediatrics* 2014;**133**:e14–22.
4. Yoshimasu K, Kiyohara C, Minami T, et al. Maternal smoking during pregnancy and offspring attention-deficit/hyperactivity disorder: a case-control study in Japan. *Atten Defici Hyperact Disord* 2009;**1**:223–31.
5. Mick E, Biederman J, Faraone SV, Sayer J, Kleinman S. Case-control study of attention-deficit hyperactivity disorder and maternal smoking, alcohol use, and drug use during pregnancy. *J Am Acad Child Adolesc Psychiatry* 2002;**41**:378–85.
6. Milberger S, Biederman J, Faraone SV, Jones J. Further evidence of an association between maternal smoking during pregnancy and attention deficit hyperactivity disorder: findings from a high-risk sample of siblings. *J Clin Child Psychol* 1998;**27**:352–8.
7. Milberger S, Biederman J, Faraone SV, Chen L, Jones J. Is maternal smoking during pregnancy a risk factor for attention deficit hyperactivity disorder in children? *Am J Psychiatry* 1996;**153**:1138.
8. Han J-Y, Kwon H-J, Ha M, et al. The effects of prenatal exposure to alcohol and environmental tobacco smoke on risk for ADHD: a large population-based study. *Psychiatry Res* 2015;**225**:164–8.
9. Zhu JL, Olsen J, Liew Z, Li J, Niclasen J, Obel C. Parental smoking during pregnancy and ADHD in children: the Danish national birth cohort. *Pediatrics* 2014;**134**:e382–e8.
10. Skoglund C, Chen Q, D'Onofrio BM, Lichtenstein P, Larsson H. Familial confounding of the association between maternal smoking during pregnancy and ADHD in offspring. *J Child Psychol Psychiatry* 2014;**55**:61–8.
11. Sagiv SK, Epstein JN, Bellinger DC, Korrick SA. Pre-and postnatal risk factors for ADHD in a nonclinical pediatric population. *J Atten Disord* 2013;**17**:47–57.
12. Ellis LC, Berg-Nielsen TS, Lydersen S, Wichstrøm L. Smoking during pregnancy and psychiatric disorders in preschoolers. *Eur Child Adolesc Psychiatry* 2012;**21**:635–44.
13. Langley K, Heron J, Smith GD, Thapar A. Maternal and paternal smoking during pregnancy and risk of ADHD symptoms in offspring: testing for intrauterine effects. *Am J Epidemiol* 2012;**176**:261–8.
14. Sciberras E, Ukoumunne OC, Efron D. Predictors of parent-reported attention-deficit/hyperactivity disorder in children aged 6–7 years: a national longitudinal study. *J Abnorm Child Psychol* 2011;**39**:1025–34.
15. Obel C, Olsen J, Henriksen TB, et al. Is maternal smoking during pregnancy a risk factor for hyperkinetic disorder? Findings from a sibling design. *Int J Epidemiol* 2011;**40**:338–45.
16. Nomura Y, Marks DJ, Halperin JM. Prenatal exposure to maternal and paternal smoking on attention deficit hyperactivity disorders symptoms and diagnosis in offspring. *J Nerv Ment Dis* 2010;**198**:672.
17. Lindblad F, Hjern A. ADHD after fetal exposure to maternal smoking. *Nicotine Tob Res* 2010;**12**:408–15.
18. Agrawal A, Scherrer JF, Grant JD, et al. The effects of maternal smoking during pregnancy on offspring outcomes. *Prev Med* 2010;**50**:13–8.

rates of comorbidity with cigarette smoking are comparable with those of other psychiatric disorders.[188–190] ADHD patients tend to start smoking at an earlier stage. The reasons for high rates of comorbidity between ADHD and smoking remain largely unknown. It has been proposed by McClernon and Kollins that both ADHD and smoking involve dysregulation of dopaminergic and nicotinic-acetylcholinergic circuits, and that genetic variations are partially involved in the high rate of comorbidity. Most of this overlap involves genes that regulate monoaminergic transmission, with a particular focus on dopaminergic system genes that include the dopamine receptor D4 (*DRD4*) and dopamine transporter (*DAT*) and genes involved in serotonin neurotransmission: the serotonin HTR1B gene and the serotonin transporter gene (5-HTT). Evidence also suggests that nicotinic receptor genes may be associated with both smoking and ADHD. However, genetic studies of the α4 subunit (see Section 9.1.2) gene (*CHRNA4*), smoking, and ADHD delivered inconclusive results.[191]

In addition to common genetic denominators, the relationship between smoking and ADHD can be explained by the impact of nicotine on neurotransmitter systems. The dopaminergic neurotransmitting system is altered in ADHD patients, so that they exhibit lower levels of dopamine. Reduction in dopamine due to an elevated number of dopamine transporters in people with ADHD was first described as characteristic of this disorder,[192] but careful meta-analysis has shown that the density of dopamine transporters in ADHD brains is dynamic and changes depending on the previous stimulus.[193] Current hypotheses involve the mesocorticolimbic dopamine pathway and the locus coeruleus-noradrenergic system.[194] Additional abnormalities in serotoninergic, glutamatergic, or cholinergic pathways have been suggested as well.[195–197] Nicotine has been shown to stimulate dopamine release in the striatum of both animals and smokers.[198,199] Stimulation of the cholinergic system by nicotine, the cholinergic modulation of dopaminergic systems, and dopamine-mediated functions have also been proposed as possible explanations for the increased rewarding experience of smoking in ADHD patients compared with controls.[200]

6.2.2 Nicotine and ADHD: Clinical Evidence

Several clinical studies on ADHD patients and nicotine have been conducted. In an acute, placebo-controlled double-blind experiment, Levin et al. examined the effects of transdermal nicotine in adults (6 smokers and 11 nonsmokers with ADHD). Smokers were deprived of cigarettes and were given a 21 mg/day nicotine skin patch for 4.5 h during a morning session. The nonsmokers were given a 7 mg/day nicotine skin patch for 4.5 h during a morning session. Significant improvements in self-rated vigor, speed of responding, concentration, and observer-rated illness severity in both subject groups and a reduction in variability of reaction time for the smokers were reported.[201] Other clinical reports of favorable nicotine effects in individuals with ADHD are in Table 2. Some adverse effects such as nausea, stomachache, itching, and dizziness were reported as well.

TABLE 2 Clinical Trials on Nicotine and ADHD[a]

Type of Study	Subject Description	Nicotine Treatment	Outcome	References
Acute, placebo-controlled double-blind crossover	ADHD adults (6 smokers and 11 nonsmokers)	Smokers—21 mg/day for 4.5 h Nonsmokers—7 mg/day for 4.5 h Active patch and placebo given 1 week apart	Clinical-rated global improvement, self-rated vigor and concentration, improved performance on chronometric measures of attention and time accuracy	1
Placebo-controlled double-blind crossover	ADHD adults (25 abstinent smokers and 27 nonsmokers)	Smokers—21 mg/day for 2 consecutive days Nonsmokers—7 mg/day for 2 consecutive days Active patch and placebo given 1 week apart	Reduction in ADHD symptoms by 8% and negative moods by 9%, and increased cardiovascular activity during first 3–6 h	2
Crossover	15 ADHD young adults (nonsmokers)	7 mg/day for 45 min Active patch and placebo given on separate days	Improved cognitive performance	3
Double placebo-controlled double-blind crossover	ADHD adolescents (13–17 years)	A 10 mg capsule of methylphenidate, 7 mg nicotine patch, and placebo Each treatment separated by at least 2 days	Positive effects on cognitive/behavioral inhibition	4
Placebo-controlled double-blind randomized pilot trial	10 ADHD adolescents (10 ± 0.8 year, 6 males and 4 females)	Transdermal nicotine patch (5 mg/16 h)	Reduction in ADHD symptoms in learning problems and hyperactivity Side effects included nausea, stomach ache, itching, and dizziness	5

[a] *ADHD, attention deficit hyperactivity disorder.*

1. Conners CK, Levin ED, Sparrow E, et al. Nicotine and attention in adult attention deficit hyperactivity disorder (ADHD). *Psychopharmacol Bull* 1996;**32**:67–74.
2. Gehricke J-G, Hong N, Whalen CK, Steinhoff K, Wigal TL. Effects of transdermal nicotine on symptoms, moods, and cardiovascular activity in the everyday lives of smokers and nonsmokers with attention-deficit/hyperactivity disorder. *Psychol Addict Behav* 2009;**23**:644.
3. Potter AS, Newhouse PA. Acute nicotine improves cognitive deficits in young adults with attention-deficit/hyperactivity disorder. *Pharmacol Biochem Behav* 2008;**88**:407–17.
4. Potter AS, Newhouse PA. Effects of acute nicotine administration on behavioral inhibition in adolescents with attention-deficit/hyperactivity disorder. *Psychopharmacology* 2004;**176**:183–94.
5. Shytle RD, Silver AA, Wilkinson BJ, Sanberg PR. A pilot controlled trial of transdermal nicotine in the treatment of attention deficit hyperactivity disorder. *World J Biol Psychiatry* 2002;**3**:150–5.

6.2.3 Nicotine and ADHD: Preclinical Evidence

To model the effects of nicotine on brain development, prenatal exposure of rat or mice has been used. While there are many similarities between the embryonal development of rodents and humans, prenatal nicotine exposure in rats or mice only models exposure during the first two trimesters of human gestation, and results are often difficult to interpret and correlate with the human situation. The first three weeks of postnatal development of rats or mice is equivalent to the third trimester of human development.[202–204] Rodent exposure models typically involve maternal nicotine administration via repeated injection, osmotic minipump, drinking water, intravenous infusion, or inhalation to simulate *in utero* exposure in the first two human trimesters.

In rodent models, measurements of locomotor activity and analysis of performance in a variety of cognitive tasks after *in utero* nicotine exposure have been used as measurements of ADHD-like symptoms. It should be noted that those measurements are usually performed in the context of other psychiatric disease models such as cognition and attention deficits in schizophrenia. While hyperactivity, following *in utero* nicotine exposure in rats, has been reported in several studies,[205–208] nicotine-induced hypoactivity was reported as well.[209–211] The observed discrepancies may be explained by different routes of nicotine administration, different doses, and different lengths of exposure period. Therefore, current studies regarding the effects of developmental nicotine exposure on locomotor activity in rats remain inconclusive.

The effects of developmental nicotine exposure on locomotor activity in the mouse appear to be more consistent and better correlated with human findings. Swiss-Webster male mice exposed via repeated maternal nicotine injections were hyperactive during adolescence.[212] Paz et al. observed hyperactivity in 60–100-day-old C57BL6/J mice of both sexes following developmental nicotine exposure via maternal exposure through drinking water.[213] This was consistent with an earlier study, which showed that both 40- and 60-day-old male C57BL6/J mice were hyperactive after nicotine exposure via the same route.[214] Interestingly, developmentally exposed female C57BL6/J mice in this study exhibited hypoactivity at 20 days of age, though this effect was not apparent at 40 or 60 days of age.

A study performed in the early 1970s evaluated the performance of developmentally exposed rats (either daily nicotine injections or hypoxic episodes throughout gestation) and observed the impairment in performing fixed ratio, variable interval discrimination, and discrimination reversal schedule appetitive tasks, which are suggestive of aberrantly rigid cognitive processing.[215] Developmental exposure (6 mg/kg/day of nicotine in drinking water of pregnant rats) also induces deficits in radial-arm maze performance, indicating deficits in learning and attentional control.[216] Furthermore, when Long-Evans rats were chronically exposed to nicotine, prenatally and postnatally via subcutaneous infusions (0.96 mg/kg/day) and pups via maternal milk until postnatal day 11, impaired spatial reference memory and mild deficits in spatial learning were observed in nicotine-exposed females in the Morris water maze.[204] In mice (HS/Ibg strain), developmental nicotine exposure impairs performance in both the radial-arm maze and the Morris water maze, which is consistent with findings from rat studies.[217] Similarly to locomotor tests in rodents, the dose, route of administration, duration, and timing of exposure are important variables influencing the appearance of deficits in cognitive tasks, as some studies reported no significant exposure effects in the radial-arm maze, Morris water maze, T-maze, or Cincinnati water maze tasks.[218–220] Additionally, nicotine exposure increased the sensitivity of cognitive functions to the effects of stress. For example, in one study, developmental nicotine exposure had little effect on radial-arm maze performance unless the test environment was altered, thereby stressing the animals.[220]

In summary, based on the data outlined above, developmental nicotine exposure can impair cognitive and attentional function in rodents, much like tobacco smoke exposure does in humans. The effect level can be variable and may be related to various factors of experimental design, making a direct comparison of study results challenging. Furthermore, the evaluation of the impact of prenatal nicotine administration may depend on the level of stress experienced during testing.

6.2.4 Summary

When comorbid with TS, prenatal maternal smoking has been reported to correlate with ADHD. High rates of comorbidity between ADHD and smoking have been described, but the etiology of this relationship is largely unknown. In addition to common genetic denominators, the relationship between smoking and ADHD can be explained by the impact of nicotine on neurotransmitter systems. The dopaminergic neurotransmission system is altered in ADHD patients, so that they exhibit lower levels of dopamine due to the elevated number of dopamine transporters. Stimulation of the cholinergic system by nicotine, cholinergic modulation of dopaminergic systems, and dopamine-mediated functions were also proposed as possible explanations for the increased rewarding experience of smoking in ADHD patients. Clinical trials with nicotine patches conducted so far reported positive results such as improved cognitive performance and overall reduction of ADHD symptoms with some adverse effects (nausea, stomach ache, itching, and dizziness). Nicotine nasal spray and transdermal patches are proposed for ADHD patients to aid in quitting smoking.

Chapter 7

Depression

7.1 SYMPTOMS, EPIDEMIOLOGY, AND POTENTIAL CAUSES

According to the World Health Organization, depression was the leading cause of disability and fourth leading contributor to the global burden of disease in 2004, with a tendency to move into first place by 2030.[221] Depression is characterized by low mood, the loss of interest, the loss of drive and pleasure, feelings of guilt, poor concentration, low self-esteem, sleep disturbances, and increased or decreased appetite.[146] Depression results from a complex interaction of social, psychological, and biological factors. People who have gone through adverse life events (unemployment, bereavement, and psychological trauma) are more likely to develop depression. In turn, depression can lead to more stress and dysfunction that worsen the affected person's life situation and exacerbate the depression itself.

A number of factors can increase a person's risk of developing depression: abuse (physical, sexual, or emotional), serious illness, substance abuse, major life events (positive and negative), genetics, and strong emotions (sadness or grief). Depression is a multifaceted disorder with diverse causes that has been associated with an increased risk of developing severe medical conditions. For instance, depression increases the risk for cardiovascular disorders, stroke, Alzemier's disease, epilepsy, diabetes, and cancer.[222]

7.2 IMPACT OF SMOKING AND NICOTINE

7.2.1 Smoking and Depression: Epidemiological Evidence

A link between smoking and depression has long been known. Smokers have higher rates of depression than nonsmokers.[21] Although the relationship between smoking and depression is complex, common genetic and environmental influences have been identified.[223–225] Additionally, there is some evidence that the high association of smoking with depression is active in both directions, that is, that there are an increased proportion of smokers among depressive patients and that there are generally more depressive patients among smokers than nonsmokers. Several studies have reported that depressed patients use smoking as self-medication.[158,226–228] Table 1 summarizes meta-analyses in which the correlation between smoking and depression was examined. Other studies are presented in Table 2.

Nicotine and Other Tobacco Compounds in Neurodegenerative and Psychiatric Diseases.

https://doi.org/10.1016/B978-0-12-812922-7.00007-X

TABLE 1 Epidemiological Studies on Cigarette Smoking and Depression—Meta-Analyses

Studies Included	Smoking Predicting Depression (OR[a]/RR[b], 95% CI[c])	Depression Predicting Smoking (OR/RR, 95% CI)	References
78 Cross-sectional studies before 2012	Current smoker 1.50 (1.39–1.60)	1.40 (1.17–1.68)	1
	Past smoker 1.21 (1.13–1.30)		
7 Cohort studies before 2012	Ever smoker 1.62 (1.10–2.40)	NA	
6 Cohort studies between 1990 and 2007	1.73 (1.32–2.40)	NA	2
12 Cohort studies between 1990 and 2007	NA	1.41 (1.21–1.63)	

[a] *OR, odds ratio.*
[b] *RR, relative risk.*
[c] *CI, confidence interval.*
1. Luger TM, Suls J, Vander Weg MW. How robust is the association between smoking and depression in adults? A meta-analysis using linear mixed-effects models. *Addict Behav* 2014;**39**:1418–29.
2. Chaiton MO, Cohen JE, O'Loughlin J, Rehm J. A systematic review of longitudinal studies on the association between depression and smoking in adolescents. *BMC Public Health* 2009;**9**:356.

7.2.2 Nicotine and Depression: Clinical Evidence

The observation that patients with depression tend to smoke has lead to the idea that nicotine may ease depressive symptoms. Estimates of the prevalence of nicotine dependence in patients with major depression range between 50% and 60%, compared with approximately 25% in the general population. Janowsky and colleagues proposed that excessive acetylcholine may lead to depression and suggested that depression is associated with hyperactivation of the cholinergic system and decreased activity of the noradrenergic system.[229] The importance of cholinergic system dysregulation in depression has been further emphasized by Marina Picciotto and her team.[230] Chronic administration of low nicotine levels is thought to desensitize nAChRs and may modulate the effects of nicotine on alleviation of depressive symptoms. It has been reported that one puff of a cigarette is enough to saturate the high-affinity nAChRs (those containing the β2 nAChR subunit) in the human brain,[231] and it is known that nicotine binding is followed by a long-term decrease in nAChR activity due to desensitization.[25] It can be hypothesized that the initial increase of nAChR activity after smoking, through upregulation of nAChRs, could lead to affective

TABLE 2 Other Epidemiological Studies on Cigarette Smoking and Depression

Type of Study	Population Size	Smoking Predicting Depression (OR[a]/RR[b], 95% CI[c])	Depression Predicting Smoking (OR/RR, 95% CI)	References
Cohort study	15,628	0.7 (0.6–0.9)	0.6 (0.4–0.8)	1
Cohort study	2101	2.3 (1.3–4.0)	NA	2
Cohort study	1770	Current smoker 2.0 (1.2–3.4)	NA	3
		Past smoker 1.2 (0.7–2.1)		

[a]*OR, odds ratio.*
[b]*RR, relative risk*
[c]*CI, confidence interval.*

1. Cheng HG, Chen S, McBride O, Phillips MR. Prospective relationship of depressive symptoms, drinking, and tobacco smoking among middle-aged and elderly community-dwelling adults: results from the China Health and Retirement Longitudinal Study (CHARLS). *J Affect Disord* 2016;**195**:136–43.
2. Bakhshaie J, Zvolensky MJ, Goodwin RD. Cigarette smoking and the onset and persistence of depression among adults in the United States: 1994–2005. *Compr Psychiatry* 2015;**60**:142–8.
3. Goodwin RD, Prescott M, Tamburrino M, Calabrese JR, Liberzon I, Galea S. Smoking is a predictor of depression onset among National Guard soldiers. *Psychiatry Res* 2013;**206**:321–3.

symptoms in depressed patients, while the long-term decrease in nAChR activity as a result of desensitization may result in the alleviation of depressive symptoms. The increase in nAChR number is maintained after quitting smoking for at least 2 weeks[232] and may contribute to depressive symptoms following abstinence. In line with the idea that decreased nAChR activity may act as an antidepressant in human smokers, studies on KO mice lacking the high-affinity subclass of nAChRs (those lacking the β2 subunit) showed decreased baseline immobility in the tail suspension and forced swim tests.[233]

Several clinical studies with nicotine patches have been conducted so far (Table 3). In general, declines in depressive symptoms were observed and reported. However, those studies were conducted on rather small numbers of subjects, and the findings should be interpreted carefully. For example, the most recent randomized, double-blind study on 20 healthy volunteers with a positive family history of depression examined the effects of transdermal nicotine on mood and electroencephalogram changes accompanying transient reductions in serotonin induced by acute tryptophan depletion (ATD). While acknowledging the limitation of the study outcome due to the small number of subjects, the incomplete assessment of family history, the employment of a mixed sample of healthy smokers and nonsmokers, and the restricted use of an acute, slowly

TABLE 3 Nicotine Trials on Depressive Symptoms

Study Design	Sample Description	Nicotine Treatment	Outcome	References
Placebo-controlled, double-blind	8 Major depressed patients and 8 healthy volunteers (all nonsmokers)	Transdermal nicotine patch (17.5 mg) or placebo for 24 h	Increased REM[a] sleep time and short-term mood improvements in patients	1
Observational	12 Major depression patients (two eventually dropped out)	Transdermal nicotine patch (17.5 mg/day) for 4 consecutive days	Improved depression but relapsed 3–4 days after last patch administration	2
Observational	12 Patients with major depression and insomnia (all nonsmokers)	Transdermal nicotine patch (17.5 mg/day)	Chronic nicotine administration improved mood and sleep	3
Placebo-controlled, double-blind	11 Depressive patients (nonsmokers)	Placebo or nicotine patches (3.5 mg/day in weeks 1 and 4 and 7 mg/day in weeks 2 and 3)	Acute nicotine did not alter mood, but chronic administration improved depression, although the effects lessened on week 4 with decreased dose	4
Randomized, double-blind	20 Healthy volunteers with a positive family history of depression	Smokers, nicotine patch (21 mg/day) Nonsmokers, nicotine patch (7 mg/day). Serotonin decrease induced by transient tryptophan depletion	Nicotine stabilized lowered mood and frontal functional asymmetry elicited by acute decrease in brain serotonin	5

[a]*REM, random eye movement.*

1. Salin-Pascual R, Galicia-Polo L, Drucker-Colín R, de la Fuente J. Effects of transdermal nicotine on mood and sleep in nonsmoking major depressed patients. *Psychopharmacology* 1995;**121**:476–9.
2. Salín-Pascual RJ, Rosas M, Jimenez-Genchi A, Rivera-Meza BL. Antidepressant effect of transdermal nicotine patches in nonsmoking patients with major depression. *J Clin Psychiatry* 1996;**57**:387–9.
3. Haro R, Drucker-Colín R. Effects of long-term administration of nicotine and fluoxetine on sleep in depressed patients. *Arch Med Res* 2004;**35**:499–506.
4. McClernon FJ, Hiott FB, Westman EC, Rose JE, Levin ED. Transdermal nicotine attenuates depression symptoms in nonsmokers: a double-blind, placebo-controlled trial. *Psychopharmacology* 2006;**189**:125–33.
5. Knott V, Thompson A, Shah D, Ilivitsky V. Neural expression of nicotine's antidepressant properties during tryptophan depletion: an EEG study in healthy volunteers at risk for depression. *Biol Psychol* 2012;**91**:190–200.

absorbed, single dose of nicotine, the authors suggest that nicotine acts to stabilize the mood-lowering and associated frontal functional asymmetry elicited by an acute decrease in brain serotonin.[234]

7.2.3 Nicotine and Depression: Preclinical Evidence

Nicotine has been tested in the preclinical setting (summarized in Table 4), with reported reductions in symptoms of depression. Additionally, nicotine agonists have also been tested in animal models for antidepressant effects. For example, a nonselective, noncompetitive nicotinic antagonist, mecamylamine, has been shown to exert antidepressant-like effects in mice in several common tests of antidepressant efficacy, including the tail suspension and forced swim tests.[230,235,236]

7.2.4 Summary

Smoking and depression are showing bidirectional association: Smoking increases the risk of depression, and depression increases the risk of smoking. Depression is associated with hyperactivation of the cholinergic system and decreased activity of the noradrenergic system. Nicotine has been used in clinical trials to evaluate whether it could provide relief to depressed patients. Chronic administration of low nicotine levels is thought to desensitize nAChRs, which is proposed as a hypothesis for the mechanism underlying potential alleviation

TABLE 4 Nicotine and Depression—Preclinical Studies

Model	Nicotine Treatment	Effect	References
FSL[a] and FRL[b] rats	Nicotine bitartrate (100 μg/mL) in tap water for 14 days	Improved performance in forced swim test	1
FSL and FRL rats	Nicotine (s.c.), acute or chronic (14 days)	Significantly improved performance in FSL but not FRL rats in the swim test	2
Learned helplessness rats	Osmotic pump (1.5 mg/kg/day)	Reduced number of failed escapes	3

[a]*FSL, Flinders sensitive line.*
[b]*FRL,* Flinders resistant line.
1. Djurić VJ, Dunn E, Overstreet DH, Dragomir A, Steiner M. Antidepressant effect of ingested nicotine in female rats of Flinders resistant and sensitive lines. *Physiol Behav* 1999;**67**:533–7.
2. Tizabi Y, Overstreet DH, Rezvani AH, et al. Antidepressant effects of nicotine in an animal model of depression. *Psychopharmacology* 1999;**142**:193–9.
3. Semba Ji, Mataki C, Yamada S, Nankai M, Toru M. Antidepressantlike effects of chronic nicotine on learned helplessness paradigm in rats. *Biol Psychiatry* 1998;**43**:389–91.

of depressive symptoms. Results from clinical studies are generally positive, showing improvements in mood and sleep. These findings are corroborated by preclinical tests of nicotine in animal models. However, those studies were conducted on rather small numbers of subjects, and the findings should be interpreted carefully. Further investigations on larger groups of patients would be needed for definitive conclusions on the usefulness of nicotine as a therapeutic agent for depression.

Chapter 8

Anxiety

8.1 SYMPTOMS, EPIDEMIOLOGY, AND POTENTIAL CAUSES

People with anxiety disorders frequently express intense, excessive, and persistent worry and fear about everyday situations.[146] There are several anxiety disorders: generalized anxiety disorder, social anxiety disorder (social phobia), specific phobias, and separation anxiety disorder. David Barlow defined anxiety as *a future-oriented mood state in which one is ready or prepared to attempt to cope with upcoming negative events*.[237] Often, anxiety is not an isolated condition and results from other medical conditions that require medical treatment. Common anxiety symptoms include feeling nervous, restless, or tense; a sense of impending danger, panic, or doom; an increased heart rate; breathing rapidly (hyperventilation); sweating; trembling; feeling weak or tired; trouble concentrating or thinking about anything other than the present worry; trouble sleeping; gastrointestinal problems; difficulty controlling worry; and an urge to avoid situations that trigger anxiety.

According to epidemiological surveys, one-third of the population is affected by an anxiety disorder at some point in life. Anxiety disorders are more common in women, and their prevalence is highest during midlife. These disorders are associated with a considerable degree of impairment, high health-care utilization, and an enormous societal economic burden.[238,239]

Anxiety disorders are complex diseases caused by a combination of genetic and environmental factors. Similar to other psychiatric diseases, predisposition to anxiety is created by genetics and family history, but generally, external stimuli trigger onset or exacerbation.[146] Genetic factors account for about 43% of variance in panic disorder and 28% in generalized anxiety disorder.[146a] For example, several gene polymorphisms have been found to correlate with anxiety: *PLXNA2*, *SERT*, *CRH*, *COMT*, and *BDNF*.[146b,146c,146d] Some of those genes impact neurotransmitters (such as serotonin and norepinephrine) and hormones that are implicated in anxiety.

Nicotine and Other Tobacco Compounds in Neurodegenerative and Psychiatric Diseases.

https://doi.org/10.1016/B978-0-12-812922-7.00008-1

8.2 IMPACT OF SMOKING AND NICOTINE

8.2.1 Smoking and Anxiety: Epidemiological Evidence

Studies on the behavior of smokers and their motivation to smoke suggest that anxiety reduction and stress relief are drivers for pursuing this habit. The opposite has also been reported, namely, the association between cigarette smoking and increased anxiety symptoms,[244] which corroborates the evidence from laboratory studies that failed to detect mood-enhancing effects of smoking or nicotine.[245] As nicotine is a potent compound, in the context of its action on different neurotransmitter systems via nAChRs, it has been implicated in various hypotheses related to the impact of smoking on anxiety.

8.2.2 Nicotine and Anxiety: Clinical and Preclinical Evidence

The literature on the effects of nicotine on anxiety in humans and animal models is complex and difficult to interpret because of mixed and inconclusive results. Table 1 summarizes clinical trials with nicotine, illustrating ambiguity in the results, while Table 2 presents preclinical studies, in which nicotine effects have been examined in animal models, also reporting ambiguous results.

TABLE 1 Clinical Trials on Nicotine and Anxiety

Study Design	Sample Description	Nicotine Treatment	Outcome	References
Randomized placebo-controlled, double-blind, crossover	33 Healthy nonsmokers undergoing 35% CO_2 challenge-induced anxiety	Pretreatment with a nicotine patch or placebo patch	No response to CO_2 challenge	1
Randomized placebo-controlled, double-blind, crossover	31 Healthy nonsmokers	Nicotine gum (2 mg) or placebo	Increased anxiety and neural activation elicited by unpleasant stimuli	2
Randomized placebo-controlled	11 OCD[a] patients	Nicotine (17.5 mg/day) or placebo patches for 5 days	Reduced anxiety assessed by the Beck Anxiety Inventory score	3

TABLE 1 Clinical Trials on Nicotine and Anxiety—cont'd

Study Design	Sample Description	Nicotine Treatment	Outcome	References
Randomized placebo-controlled	88 Posttraumatic stress disorder patients	High-nicotine yield cigarette or nicotine-depleted cigarette. Participants were subjected to anxiety priming	Ameliorated self-reported distress in anxiety-eliciting condition; enhanced autonomic reactivity to trauma cues	4

[a]*OCD, obsessive-compulsive disorder.*

1. Cosci F, Abrams K, Schruers KR, Rickelt J, Griez EJ. Effect of nicotine on 35% CO2-induced anxiety: a study in healthy volunteers. *Nicotine Tob Res* 2006;**8**:511–7.
2. Kobiella A, Ulshofer DE, Vollmert C, et al. Nicotine increases neural response to unpleasant stimuli and anxiety in non-smokers. *Addict Biol* 2011;**16**:285–95.
3. Salin-Pascual RJ, Basanez-Villa E. Changes in compulsion and anxiety symptoms with nicotine transdermal patches in non-smoking obsessive-compulsive disorder patients. *Rev Invest Clin* 2003;**55**:650–4.
4. Buckley TC, Holohan DR, Mozley SL, Walsh K, Kassel J. The effect of nicotine and attention allocation on physiological and self-report measures of induced anxiety in PTSD: a double-blind placebo-controlled trial. *Exp Clin Psychopharmacol* 2007;**15**:154–64.

TABLE 2 Nicotine and Anxiety—Preclinical Studies

Model	Nicotine Treatment	Effect	References
C57BL/6J mice	Nicotine (0.05–0.5 mg/kg, i.p.[a])	Low dose (0.05 mg/kg) supported anxiolytic behavior High dose (0.5 mg/kg) supported anxiogenic behavior	1
Wistar rats	Nicotine (0.4–0.8 mg/kg, i.p.)	0.6 and 0.8 mg/kg increased anxiety behavior 0.4 mg/kg was ineffective	2

[a]*i.p., intraperitoneally.*

1. Anderson S, Brunzell D. Anxiolytic-like and anxiogenic-like effects of nicotine are regulated via diverse action at β2* nicotinic acetylcholine receptors. *Br J Pharmacol* 2015;**172**:2864–77.
2. Zarrindast MR, Eslahi N, Rezayof A, Rostami P, Zahmatkesh M. Modulation of ventral tegmental area dopamine receptors inhibit nicotine-induced anxiogenic-like behavior in the central amygdala. *Prog Neuro-Psychopharmacol Biol Psychiatry* 2013;**41**:11–7.

8.2.3 Summary

Similarly to depression, a bidirectional association between smoking and anxiety has been observed. Patients with anxiety tend to smoke more aiming at anxiety reduction and stress relief. The opposite has been reported as well: smokers tending to be more anxious. Clinical and preclinical results on nicotine and anxiety are challenging for the interpretation because of mixed and inconclusive results. Further investigations would be needed for the generation of solid evidences regarding the impact of nicotine in this disorder and possible recommendation for the patients.

Part II

Overview of the Pharmacology of Nicotine and Other Tobacco-Derived Compounds

Chapter 9

Nicotine

Nicotine (Fig. 1) is a naturally occurring alkaloid found in the Solanaceae (nightshade) family and in other plants. It is produced in concentrations high enough to exert a pharmacological effect by the *Nicotiana* subfamily (up to 14% of dry weight in *N. rustica*) and *Duboisia hopwoodii* (ca. 2% of dry weight).[246] Nicotine ($C_{10}N_{14}H_2$, molecular mass 162.23), also called 3-(1-methyl-2-pyrrolidinyl) pyridine according to the IUPAC[a] nomenclature, is a bicyclic compound with a pyridine and a pyrrolidine cycle. Commercially available nicotine is an oily, colorless to light yellow liquid that is readily soluble in ethanol.[247]

9.1 PHARMACOLOGICAL CONSIDERATIONS

9.1.1 Neuropharmacological Effects of Nicotine

Nicotine has been reported to restore cognitive function in nicotine-deprived smokers and to enhance cognition in nonsmokers and experimental animals.[248,249] Furthermore, positive effects of nicotine have been demonstrated on working memory,[250] attention,[251] and recognition memory[252] in rats and short-term (working) memory[253,254] and multiple domains of cognition in humans, including attention, information processing, and short-term memory.[255–258] A meta-analysis of 41 double-blind, placebo-controlled laboratory studies published from 1994 to 2008 further confirmed these observations. Nicotine or smoking improved six domains: fine motor abilities, alerting attention-accuracy and response time (RT), orienting attention-RT, short-term episodic memory-accuracy, and working memory-RT.[259]

9.1.2 Nicotinic Acetylcholine Receptors

The molecular mechanisms of the neurological effects of nicotine are not fully understood. This alkaloid acts as a ligand for nicotinic acetylcholine receptors (nAChR), a member of the cys-loop receptor superfamily that also includes serotonin (5-hydroxytryptamine, 5-HT), specifically 5-HT3 receptors, γ-aminobutyric acid (GABA) A and C receptors, and glycine receptors and participates in a variety of physiological functions, including regulation of neuronal

a. International Union of Pure and Applied Chemistry.

Nicotine and Other Tobacco Compounds in Neurodegenerative and Psychiatric Diseases.

https://doi.org/10.1016/B978-0-12-812922-7.00009-3

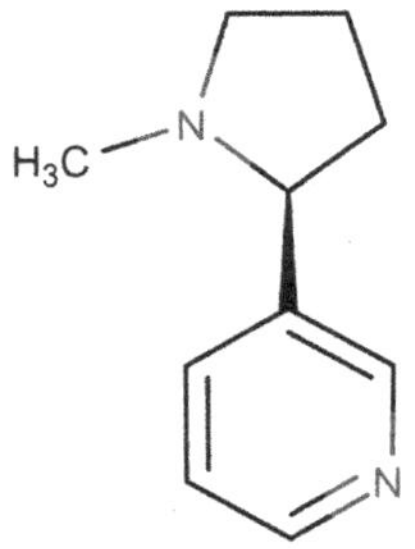

FIG. 1 Chemical structure of nicotine.

excitability and neurotransmitter release.[154,260–262] Nicotinic AChRs are present in the central and peripheral nervous systems and in the immune system and various peripheral tissues.[263,264] They are pentamers that comprise of different combinations of α1–α10, β1–β4, γ, δ, and ε subunits, with each subunit having four putative transmembrane-spanning domains and a similar topological structure. Mammalian neuronal subunits are divided into alpha (α2–α7, α9, and α10) and beta (β2–β4) based on the presence of adjacent cysteine groups in the extracellular domain of only α subunits. α7 and α9 subunits form homomeric receptors, while heteromeric nAChRs can be composed of various subunit combinations of α1–α6 and β1–β4.[265]

Differences in subunit composition and tissue localization determine the functional and pharmacological characteristics of the receptors. For example, homomeric channels composed of α7 or α9 subunits and heteromeric nAChRs containing α9 and α10 subunits show the greatest calcium permeability.[266,267] Heteromeric α4/β2 and homomeric α7 nAChRs are the most predominant subtypes found in the mammalian brain and the most commonly targeted in drug discovery programs for neurodegenerative and psychiatric disorders. Other nAChRs have a much more restricted distribution in the brain. Another interesting feature of heteromeric α4/β2 is that the pentamer can exist in two stoichiometric variants, 2 α4 and 3 β2 or 3 α4 and 2 β2, with slightly different physiological properties, such as opening, closing, desensitization kinetics (see below), ionic conductance, cationic selectivity, and pharmacological properties.[268]

Nicotinic AChRs function as ion channels and upon ligand binding allow sodium or calcium influx and potassium efflux. Most of the potential therapeutic effects of nAChR ligands have been attributed to their ability to influence neuronal excitability, synaptic plasticity, and gene expression via intracellular calcium regulation. ACh is the endogenous ligand of nAChRs. However, nicotine binds to nAChRs with a much higher affinity and longer receptor depolarization than ACh. During this period, the receptor is not responsive. This reversible reduction in response during sustained agonist application, termed receptor desensitization,[269,270] indicates that nicotine can actually act as an agonist and antagonist of nAChRs. Neuroadaptation or tolerance develops with repeated exposure to nicotine and receptor desensitization. Further, the temporary absence of responsive receptors is compensated for by upregulation of receptors and an increased number of binding sites.[26] Moreover, α4/β2 nAChRs have a higher affinity for ACh and nicotine, and also desensitize slowly, compared with α7 homopentamers, which desensitize rapidly.[262]

9.1.3 Nicotine Effects on Neurotransmitter Systems

When inhaled with cigarette smoke, nicotine rapidly reaches the brain, where a variety of neurotransmitters are released.[26] Nicotinic receptor activation/inactivation often involves not only ACh but also dopamine, noradrenaline, serotonin, GABA, and glutamate, indicating the importance of nAChRs in various neurotransmitter systems.[271] These neurotransmitter systems play an important role in cognitive and noncognitive functions such as learning, memory, attention, locomotion, motivation, reward, reinforcement, and anxiety. For example, in 1979, Giorguieff-Chesselet et al. showed that nicotine increases the levels of dopamine, a neurotransmitter essential for boosting attention, reward-seeking behavior, and risk of addiction ranging from gambling to drugs use. Dopamine also helps to control movement. Nicotine receptors in the striatum are located near the terminals that regulate and release dopamine. Small nicotine doses stimulate striatal dopamine release and thereby control movement that would otherwise go uncontrolled.[198] *In vitro*, nicotine protection against glutamate toxicity could be blocked by dopamine-1 receptor antagonists in retinal neurons, suggesting that this neuroprotection is mediated by stimulation of dopamine release.[272] Indirect modulation of dopamine release by nicotine is exerted via nAChRs located on excitatory glutamatergic and inhibitory GABAergic neurons in the ventral tegmental area. Binding of nicotine to nAChRs located on excitatory glutamatergic terminals results in glutamate release, which in turn stimulates dopaminergic neurons. Binding of nicotine to nAChRs located on inhibitory GABAergic projections leads to the release of GABA, which in turn inhibits dopaminergic neurons. The relevance of nicotine's impact on levels of dopamine in the context of Parkinson's disease (PD) is particularly interesting, since the loss of dopaminergic neurons is a characteristic of PD. The impact of nicotine on the cholinergic system by upregulation of the expression of nAChRs is of particular importance in Alzheimer's disease (AD), since the loss of cholinergic neurons is a characteristic of AD. The importance of nicotine and its impact on other neurotransmitter systems in the context of other psychiatric diseases and addiction have been reviewed in more details in several articles.[25,273,274]

9.1.4 Nicotine Effects on Molecular Signaling

The mode by which nAChRs interact with signal transduction molecules varies depending on the type of neuronal and nonneuronal cells targeted by nicotine and the nAChR subunits involved. Activation of nAChRs results in channel opening and increased intracellular calcium concentration, which affects downstream molecular signaling and is critical for neuroprotection in target cells. Calcium entry can occur directly through nAChRs[275,276] and particularly α7 nAChRs, which exhibit high calcium conductance.[277] Additionally, calcium levels can be indirectly increased via activation of voltage-gated calcium channels or calcium release from intracellular stores.[278–280] Changes in intracellular calcium concentration affect various calcium-dependent molecular cascades and cell processes.[281]

Importantly, calcium-independent nAChR-mediated mechanisms have also been reported,[282] suggesting that the molecular events triggered by nicotine are a combination of ion-dependent and ion-independent pathway activation/inactivation, with each type of signaling necessary for specific biological responses. For example, it is unclear whether α7 receptors in nonneuronal cells are capable of ion channel activation, but it has been clearly demonstrated that they mediate other forms of signal transduction.[23] Additionally, binding of nicotine agonists causes conformational changes in nAChRs that are dependent on the specific binding position and that may variously promote activation or desensitization or alternatively function as competitive antagonists.[283] In other words, it is possible that desensitized (nonconducting) receptors, together with activated (conducting) receptors, confer the resultant effects, which adds to the overall complexity of nAChR physiology.

Downstream signaling triggered by activated nAChRs and calcium ion influx[72] includes a number of molecules: phospholipase C, protein kinase C (PKC) isoforms, phosphatidylinositol 3-kinase (PI3K), Akt or protein kinase B (PKB), c-Jun N-terminal kinase (JNK), Src, Janus kinase 2 (JAK2), p38 mitogen-activated protein kinase (MAPK), and RAS/RAF/MEK/ERK pathway.[284] Upon nAChR activation, these signaling molecules can be both activated and inactivated (phosphorylated and dephosphorylated) owing to conformational changes of nAChR subunits and/or associated proteins. Some of these signaling pathways are involved in nicotine-mediated neuronal protection. Kihara et al. reported nicotine-mediated neuronal protection against Aβ neurotoxicity in an AD model of primary cortical neurons. Protection was mediated via physical association between α7 nAChRs and PI3K, with nicotine-increasing activity of Akt, a PI3K effector. Expression of B-cell lymphoma 2 (Bcl-2), a cell survival protein, was also increased by nicotine in the same study.[285] Further, nicotine decreases Aβ-induced caspase-3 activation in hippocampal neurons,[286] decreases activation of the downstream cell death effector JNK in tumor cells,[287] and decreases cytochrome *c* release while activating caspase-3, caspase-8, caspase-9, DNA laddering, and neuronal NO synthase activity in spinal cord cultures exposed to arachidonic acid.[288,289]

Dajas-Bailador et al. reported that nicotine stimulation (100 μM) leads to protein kinase A (PKA) activation via α7 nAChRs and the downstream RAF/MEK/ERK signaling pathway in human SH-SY5Y neuroblastoma cells and hippocampal neurons and may be responsible for cell survival.[290] Other groups have provided evidence that nicotine stimulation of α7 nAChRs transduces signals to PI3K via a JAK2 cascade, which results in neuroprotection against Aβ-induced apoptosis in PC12 (rat pheochromocytoma) cells.[291] Furthermore, stimulation of α7 nAChRs by either their physiological ligand ACh or nicotine in human oral keratinocytes leads to altered gene expression due to transactivation of signal transducer and activator of transcription 3 (STAT3), which occurs through two complementary signaling pathways coupled by the receptor. The pathway mediated by stepwise activation of RAS/RAF/MEK/ERK provides increased cytoplasmic concentration of STAT3 due to upregulated expression, whereas activation of the tyrosine kinase, JAK2, causes STAT3 phosphorylation with subsequent nuclear translocation of STAT3 dimers to alter gene expression and activate the JAK2/STAT3 pathway.[292]

De Jonge et al. showed that phosphorylated STAT3 is involved in antiinflammatory α7 nAChR-dependent signaling in peritoneal macrophages.[293] Nicotine acts on macrophages by JAK2 recruitment to α7 nAChRs. Activation of JAK2 initiates the antiinflammatory STAT3 and suppressor of cytokine signaling 3 (SOCS3) signaling cascade, possibly through a direct interaction between dimerized STAT3 and the p65 subunit, which leads to inhibition of nuclear factor κB (NF-κB) p65 transcriptional activity. Another study also found that nicotinic stimulation of α7 receptors suppresses NF-κB signaling.[294] Hosur and colleagues focused their research on α4β2-mediated effects of nicotine and reported that nicotine suppresses constitutive NF-κB activity and proinflammatory cytokine production.[295] Further, they described a calcium and cAMP-PKA-independent signaling cascade induced by nicotine and suggested a role for JAK2/STAT3 transduction in α4β2-mediated attenuation of lipopolysaccharide (LPS)-induced inflammation in human neuroblastoma cells.[282]

Another signal transduction pathway activated by nicotine involves calmodulin, calcium effector protein, and PI3K-/Akt-dependent signaling. Toulorge et al. suggested that slight but chronic elevations in cytoplasmic calcium concentration in midbrain dopamine neurons *in vitro*, induced by nicotine and concomitant depolarizing treatments, activate a survival pathway involving calmodulin and PI3K.[296]

Activation of these diverse signaling cascades (Fig. 2) is reported to modulate caspase activity (3, 8, and 9), cell survival proteins (such as Bcl-2 and Bcl-x), NF-κB, CREB, tyrosine hydroxylase, and other molecular components.[13,282,297–300] Such profound molecular effects may decrease apoptosis and enhance cell survival, modify immune responses, and cause synaptic plasticity alterations. The role of microglial α7 nAChRs in calcium influx-mediated upregulation of Aβ phagocytosis via actin reorganization has been identified as an additional neuroprotective mechanism.[300]

Nicotine also causes changes in various growth factors in neuronal tissue, which may attenuate neuronal damage. Nicotine induces basic fibroblast growth factor-2 (bFGF-2) via stimulation of α4β2 nAChRs in the cortex, hippocampus, substantia nigra, and striatum of rats, further supporting its role in neuroprotection.[301–304] Additionally, the neurotrophin receptor,

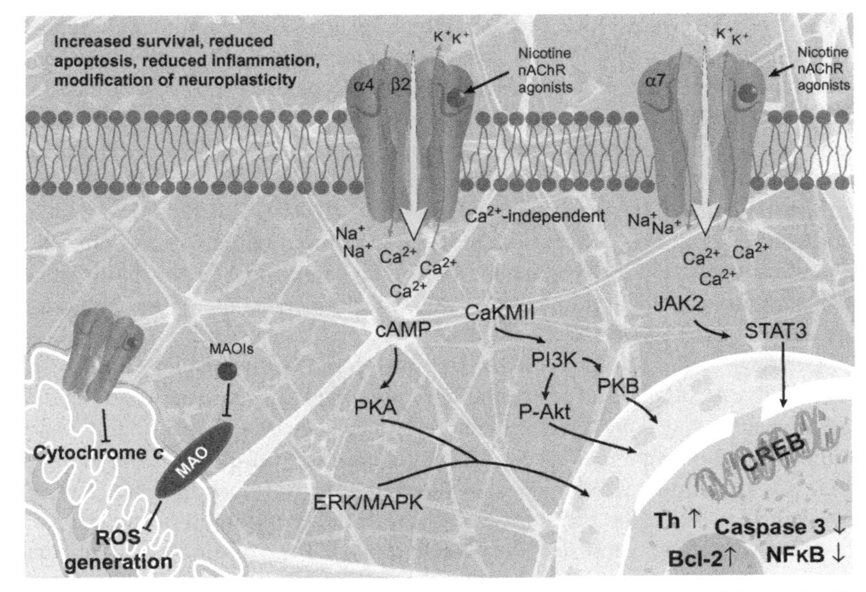

FIG. 2 Putative nicotinic acetylcholine receptor (nAChR)-mediated intracellular molecular mechanisms involved in neuronal protection. Nicotine-induced increase in intracellular calcium levels activates protein kinase A (PKA) and extracellular signal-regulated mitogen-activated protein kinase (ERK/MAPK), calcium effector protein calmodulin (CaKMII) and phosphatidylinositol 3-kinase (PI3K)/serine threonine protein kinase (Akt) or protein kinase B (PKB)-dependent signaling, and JAK2 (Janus kinase 2)/PI3K and/or JAK2/STAT3 (signal transducer and activator of transcription 3) pathways. Activation of those pathways results in modulation of caspase activity; cell survival proteins such as B-cell lymphoma 2 (Bcl-2), nuclear factor kappa B (NFκB), cAMP response element-binding (CREB), and tyrosine hydroxylase (Th); and other molecular mechanisms, leading to decreased apoptosis, enhanced neuronal survival, modified immune responsiveness, and alterations in synaptic plasticity. *ROS*, reactive oxygen species; *MAO*, monoamine oxidase; *MAOIs*, monoamine oxidase inhibitors. *This illustration was made based on findings from two publications: Quik M, Perez XA, Bordia T. Nicotine as a potential neuroprotective agent for Parkinson's disease. Mov Disord 2012;27:947–57; Gergalova G, Lykhmus O, Kalashnyk O, et al. Mitochondria express α7 nicotinic acetylcholine receptors to regulate Ca2+ accumulation and cytochrome c release: study on isolated mitochondria. PLoS One 2012;7:e31361.*

tyrosine receptor kinase A (trkA), is upregulated in both PC12 cells[305] and hippocampus[306] via α7 receptor activation. Nerve growth factor (NGF) and its receptor, tropomyosin receptor kinase B (trkB), are upregulated in the hippocampus.[307] Finally, several transcription factors that act downstream of growth factor receptors are also upregulated following repeated nicotine administration, including nuclear orphan receptor 77 (Nur77) and early growth response 1 and 2 (Egr-1 and Egr-2).[308] Taken together, these findings suggest that both bFGF-2 and NGF are candidate effectors of nicotine-mediated neuroprotection.[309–312]

Nicotine may also act by modulating cytokine production, protecting against toxic insults, and promoting neuronal repair.[130,282,314,315]

9.1.5 nAChRs in Mitochondria

Many neurodegenerative diseases demonstrate abnormal mitochondrial morphology and biochemical dysfunction. Indeed, mitochondrial dysfunction is part of an even more complex, yet poorly understood neurodegenerative pathophysiology that includes inflammation, oxidative stress, apoptosis, autophagy, and other noncausally linked biological processes.[316] Activity of mitochondrial complex I[b], II[c], and IV[d], which is critical for optimal functioning of the electron transport system, is reduced in the substantia nigra of PD patients.[317–319] Additionally, Aβ progressively accumulates in neuronal mitochondria from human AD patients and from AD model mice.[320,321]

Nicotine actions that may lead to neural protection include modification of mitochondrial complex I activity, inhibition of ROS generation, oxidative or antioxidative potential, and radical scavenging properties.[22,300,322–326] Interestingly, Maryna Skok's research group identified α7, α7β2, α4β2, α3β2, and α3β4 nAChRs in the mouse liver, brain, and lung mitochondria, a finding that contradicts the established dogma of purely plasma membrane-located functional nAChRs.[313,327] Using selective ligands for different nicotinic receptor subtypes and apoptogenic agents that trigger different mitochondrial signaling pathways,[328] α7β2 receptors were found to mainly affect the mitochondrial PI3K/Akt pathway in the outer mitochondrial membrane extracts from the mouse liver. However, α3β2 and α4β2 nAChRs significantly influenced calcium-calmodulin-dependent kinase II (CaMKII) and Src (proto-oncogene, nonreceptor tyrosine kinase)-dependent pathways in mitochondria (Fig. 3.). Thus, it was proposed that nAChRs within the mitochondrial membrane are involved in control of apoptosis. The synergistic effects of nicotine action on the cell membrane and the mitochondrial nAChRs may result in a combination of growth-promoting and antiapoptotic signals, which have recently been proposed by Chernyavsky et al. as a tumor-promoting mechanism in lung cells.[329] That study used lung epithelial cell lines (lung tumor, normal, and transformed) to demonstrate interactions between specific nAChR subunits on the cell membrane and growth factors. The activation of α7 nAChRs mainly synergized with epidermal growth factor (EGF), α3 with vascular endothelial growth factor (VEGF), and α4 with insulin-like growth factor I (IGF-I) and VEGF. α9 synergized with EGF, IGF-I, and VEGF, and also with a reduction in mitochondria-mediated apoptosis (measured by formation of the mitochondrial

b. NADH-quinone oxidoreductase catalysis, the first step in the electron transport chain of human mitochondria.

c. Succinate dehydrogenase of Krebs cycle, coupling two-electron oxidation from succinate to fumarate.

d. Cytochrome *c* oxidase, contributing to establishment of the proton gradient.

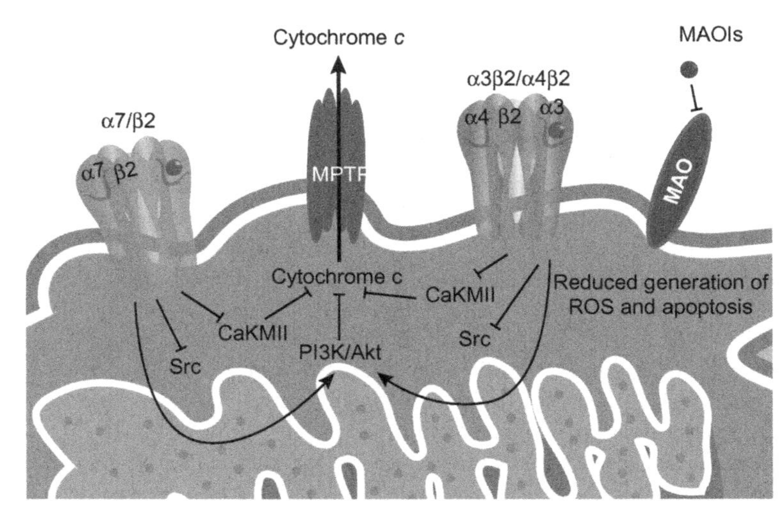

FIG. 3 Proposed mitochondrial nicotinic acetylcholine receptor (nAChR)-mediated signaling. Activated nAChRs activate the phosphatidylinositol 3 kinase (PI3K)/serine threonine protein kinase (Akt) pathway and inhibit calcium-calmodulin-dependent kinase II (CaKMII) and proto-oncogene, nonreceptor tyrosine kinase (Src)-related pathways, leading to inhibition of cytochrome *c* release. *ROS*, reactive oxygen species; *MAO*, monoamine oxidase; *MAOIs*, monoamine oxidase inhibitors. *This illustration was made based on findings by Lykhmus O, Gergalova G, Koval L, Zhmak M, Komisarenko S, Skok M. Mitochondria express several nicotinic acetylcholine receptor subtypes to control various pathways of apoptosis induction. Int J Biochem Cell Biol 2014;53:246–52.*

permeability transition pore). It is possible that the same molecular pathways are involved in cellular protection and survival in neuronal tissue following nicotine exposure.

9.1.6 nAChRs in Neural Stem Cells

The discovery of neural stem cells changed the prevailing dogma that the nervous system lacks regenerative power. Neural stem cells have the capacity to self-renew and generate the three major cell types of the CNS, offering new avenues for the treatment of neurodegenerative disorders. Not surprisingly, neural progenitor cells (NPCs) express nAChRs.[330–332] Examination of mRNA expression in rat and mouse neurocortex (18-day-old embryonic Wistar rats or 15.5-day-old embryonic Std-ddY mice) identified α2–α5, α7, α9, β2, and β4 nAChR subunits in rat and α3–α5, α7, α9, β2, and β4 nAChR subunits in murine neuronal progenitor cells. Proteins were detected for α4 and β2 subunits in the mouse brain.[330] Fetal mouse cerebral cortical neural precursors express mRNA for α3, α4, α7, β2, and β4 nAChR subunits.[333] Takarada et al. further characterized the potential role of nAChRs identified in cultured neuronal progenitor cells from fetal mouse neocortex and suggested a dual role for α4β2 activation by nicotine through a mechanism involving extracellular calcium ion influx:

proliferation toward self-replication and commitment/differentiation into particular progeny lineages in neural progenitors during brain development.

The presence of functional nAChRs in neural stem cells provides a promising research line for neural stem-cell fate and therapeutic intervention in neuroinflammatory diseases.

9.1.7 nAChRs in Microglia

Depending on the state of activation and release of mediators, microglia perform both neuroprotective and neurotoxic functions in the brain. While microglia-driven neuroinflammation has a beneficial effect on scavenging cell debris, tissue healing, and repair, chronic activation of this cell type leads to harmful neuronal effects, which contribute to the pathophysiology of neurodegenerative diseases. Although nAChR function in neuronal populations is well characterized, nAChR expression and relevance in glial cells has not yet been thoroughly characterized. Shytle et al. described for the first time the existence of a brain cholinergic pathway that regulates microglial activation via α7 nAChRs and observed that nicotine pretreatment prevents LPS-induced tumor necrosis factor-alpha (TNF-α) release in murine-derived microglial cells.[141] In rat microglial cultures, nicotine activates α7 nAChRs, inducing the synthesis of mediators involved in immunomodulation (cyclooxygenase 2 (COX-2) and prostaglandin E2), which may be part of an antiinflammatory pathway regulated by the cholinergic system.[334] The antiinflammatory role of activated α7 nAChRs may not be derived from channel opening. When the receptor is blocked by α7 antagonists (e.g., methyllycaconitine) or partial agonists (e.g., GTS-21, a derivative of anabaseine), LPS-induced TNF-α release is reduced, suggesting that ion flux through the receptor is not responsible for the reduction in TNF-α release.[335]

In AD, astrocytes are implicated in the formation of senile plaques, one of the core pathological features of the disease. As extensively reviewed by Sadigh-Eteghad and colleagues, there is evidence that glial nAChR dysfunction may contribute to neurotransmission disruption in AD.[336] An increase in astrocytic α7 but not α4 nAChR subunits in the hippocampus and entorhinal cortex has been observed in AD patients compared with age-matched controls.[337] Those findings corroborate results from animal studies with AD models, which suggest an involvement of astrocytic α7 nAChRs in the metabolism of Aβ and a connection with the amyloid cascade. A neuroprotective effect for cholinergic microglial stimulation during an ischemic event was observed in organotypic hippocampal cultures subjected to oxygen and glucose deprivation (*in vitro*) and in photothrombotic stroke (*in vivo*). The results of these experiments suggest that activation of the microglial α7 nAChR/Nrf2/heme oxygenase 1 (HO-1) axis regulates neuroinflammation and oxidative stress, providing neuroprotection under brain ischemic conditions.[338] Furthermore, it has been shown that overactivation of astrocytic α4β2 and α7 nAChRs promotes the expression of glial cell-derived neurotrophic factor (GDNF).[339,340] Astrocyte-derived GDNF binds to GDNF family receptor alpha-1 (GFRa1) and activates intracellular signaling cascades responsible for inhibiting microglial activation, protecting neurons from neurodegeneration.[341]

9.1.8 Transmembrane Nicotine Transport

Transport of nicotine from the blood to the brain is required for exerting its neurobiological effects. In human subjects, inhaled nicotine reaches 50% of its maximum brain concentration within 15 s.[342] After reaching the CNS, nicotine acts on nAChRs and can cross cellular membranes. Physicochemical properties of nicotine solutions indicate that 19% of nicotine is in its nonprotonated form at pH 7.4[343,344] and is assumed to cross the blood-brain barrier (BBB) by passive diffusion (which is fast enough to produce its effects), whereas the remaining 81% of protonated nicotine would require an active transport system to enter cells.[345] So far, the nicotine transport system has been characterized in multiple cell types, but the transporter or transporters involved have not been identified and characterized.

Coexisting passive and active nicotine transport (facilitated antiporter-mediated flux) was first characterized in Caco-2 cells and in renal basolateral membranes.[346,347]

Active and passive nicotine transport in the mouse luminal BBB was investigated by *in situ* brain perfusion by Cisterino et al. Nicotine influx followed Michaelis-Menten kinetics (K_M=2.60 mM and V_{max}=37.60 nmol/s/g at pH 7.40). Most nicotine transport (79%) was mediated by an active transporter, while passive diffusion accounted for 21%. Experiments with KO mice for characterized transports and chemical inhibition of other transporters did not alter nicotine transport at the BBB. The *in vivo* manipulation of intracellular and/or extracellular pH, the chemical inhibition profile, and the trans-stimulation experiments demonstrated that the nicotine transporter at the BBB shared the properties of the clonidine/proton antiporter.[348] Nicotine transport in the rat brain and liver has been characterized as well.[349,350] Recently, characterization of the transporter system was thoroughly performed in primary rat lung epithelial cells. A novel proton-coupled antiporter is involved in the uptake of nicotine in alveolar epithelial cells and its absorption from the lungs into the systemic circulation, but the transporter itself has not been identified and biochemically characterized. The authors suggest that this newly characterized transporter may be active in brain capillary endothelial cells, and retinal capillary endothelial cells may be similar or the same.[351] Nicotine transport has been studied in other cell types, such as the LLC-PK1 kidney epithelial cell line[352] and the JAR human choriocarcinoma cell line.[353]

9.1.9 nAChRs in Developing Brain

nAChRs are present in the brain very early in development and exhibit different dynamics of expression in different brain structures. Fluktuations in expression occur during critical periods of neuronal development suggesting their prominent role in various biological processes important for neuronal development.

α7

nAChRs are widely distributed throughout the human and rodent brain during all phases of development.[354–356] Already in the first trimester of embryonal development in humans, both α7 mRNA and protein have been detected in the fetal brain and spinal cord, just before the onset of synaptogenesis. Throughout prenatal development an increase in the expression of this receptor has been reported.[357,358] Similarly, α7 mRNA is first detected in the rodent cortex already in gestational day 13 and then shows a transient peak at around 1 week postnatal and then declines to adult levels.[359–362] The expression pattern is not constant over time; it starts homogenously across cortical regions, but during the early postnatal period, the patterns of α7 mRNA and protein expression become heterogeneous: sensory and limbic cortices showing elevated mRNA and protein levels compared with surrounding areas with the transient expression of α7 nAChRs delineating the sensory cortical regions with a spatiotemporal correspondence to ingrowing thalamic afferents and is tightly regulated by thalamic afferent activity.[362] Early and transient expression of α7 nAChRs has also been observed in other brain regions and spinal cord of rodents[363,364] suggesting an important role for this nAChR subtype in various mechanisms of neuronal development.[355]

α4β2

During early brain development, α4β2 nAChRs are widely distributed in humans. mRNA of α4 and β2 nAChR subunit have been detected in the fetal cortex, hippocampal formation, spinal cord, and other brain regions as early as the first trimester.[357,365] Similarly, early expression of α4 and β2 nAChR subunit mRNAs and high-affinity nicotine binding sites have been detected in the developing rat CNS as of gestational days 12–13, and detectable expression at different levels in brain regions is maintained throughout the prenatal period.[354,366] After birth, the expression levels of these mRNA transcripts range: slow decline in some regions to increasing levels in others after the first postnatal week and into adulthood.[367,368]

α3β4

Both α3 and β4 transcripts are widely distributed in the rat brain throughout midgestation (day 13) but decline at later developmental stages.[354,369] Region specific distribution has been observed. In the hippocampus and neocortex, the expression of α3 and β4 is transiently increased in the early postnatal period and then declines to levels found in the adult. In the other regions of the brain, stable expression of α3 and β4 mRNA transcripts can be seen throughout postnatal development and into adulthood. The expression of α3 only declined during gestation, and greater correspondence in the temporal and spatial expression of α3 and β4 subunit mRNAs is evident. This suggests that α3β4 subunits may be coordinately regulated and may form functional neuronal nAChRs with significant developmental roles.[369]

9.1.10 Summary

The impact of nicotine on cognitive functions is well-known and has been a topic of interest for a long time. Nicotine acts on nicotinic receptors, a multisubunit membrane complex that functions as ion channels, allowing sodium or calcium entry and potassium efflux upon ligand binding. In the brain, nicotine induces release of a number of neurotransmitters such as ACh, dopamine, noradrenaline, serotonin, GABA, and glutamate; this release is central to its function in the context of neuroprotection. Activated receptors trigger several calcium-dependent and -independent molecular signaling cascades. The following molecular pathways may be critical for neuroprotection: Activation of antiapoptotic genes, reduced inflammation, activation of prosurvival genes, and modification of neuroplasticity. The existence of a variety of subunit combinations with specific functions expressed in the CNS and the periphery, combined with the fact that they can be activated and desensitized, provides a number of multimodular functionalities that could perhaps be leveraged in the development of therapeutic agents.

9.2 IMPACT OF NICOTINE ON BRAIN DEVELOPMENT

Prenatal nicotine exposure, most often via tobacco smoke has been shown to have concerning effects on human brain development. The effects of prenatal nicotine exposure are broad, affecting both behavioral and neuronal development in several brain regions. The molecular mechanisms by which nicotine exerts these effects are largely unclear. *In utero* exposure has been reported to be a risk factor throughout childhood and adolescence for cognitive and neurobehavioral challenges, such as ADHD.[37,38] The cholinergic system is an important neurotransmitter system for brain maturation from the earliest stages of formation through adolescence.[370] Various nAChRs, with particular physiological roles, are transiently expressed during those processes and nicotine can influence them via activation and desensitization depending on the timing, amount, and duration of exposure.

9.2.1 Impact of Prenatal Nicotine Exposure on Development of Brain

Slotkin et al. examined the effects of a continuous 16-day gestational exposure to nicotine on brain development in the offspring of Sprague-Dawley rats (6 mg/kg/day, minipump on the fourth day of gestation). This treatment resulted in lower amounts of total brain DNA content during the late gestational and early postnatal period when compared with controls. This decrease in DNA content reflects a reduction in total cell number, and suggests that nicotine inhibits or slows down the increases in cell number that characterize early brain maturation. After birth, brain development in the nicotine-exposed animals showed persistent abnormalities in the timing of maturational events, with elevated levels of ornithine decarboxylase (ODC), an enzyme critical to the coordination of biosynthesis during cell replication and differentiation, detectable in all brain regions. Because elevated ODC activity is indicative of abnormal cellular maturation, it has been suggested that early stimulation of nAChRs by exogenous nicotine may trigger an aberrant switch from neuronal proliferation to differentiation.[371] Similarly, designed study with lower nicotine doses (2 mg/kg) has been conducted by Navarro and colleagues. Small but persistent increase in [3H]nicotine binding sites in midbrain and brainstem and an elevation of ODC activity after birth in three brain regions (cerebellum, cerebral cortex and midbrain, and brainstem) that resolved by the end of the second postnatal week were observed. The total DNA amount in the cerebellum was slightly reduced, and the development of peripheral noradrenergic projections (assessed by kidney norepinephrine levels) was impaired.[372]

In the study conducted by Abdel-Rahman A et al., timed-pregnant Sprague-Dawley rats were treated with nicotine (3.3 mg/kg, via mini osmotic pump) from gestational days 4–20. Maternal exposure to nicotine produced significant neurobehavioral deficits: beam-walk time and forepaw grip time showed significant impairments in both male and female offspring. A decrease in surviving Purkinje neurons in the cerebellum and of the survival of neurons in the CA1 subfield of hippocampus on postnatal day 30 and 60 was observed. Furthermore, an increased expression of glial fibrillary acidic protein (reported to play an important role in the long-term maintenance of brain cytoarchitecture[373]) in cerebellum and CA1 subfield of hippocampus of the offspring on postnatal days 30 and 60 was detected. These neurobehavioral and pathological deficits indicate that maternal nicotine exposure may produce long-term adverse health effects in the offspring.[374] Further evidences about reduction of neuronal area and cell size in hippocampus, suggesting decrease in its function, are reported.[375,376]

In summary, animal models indicate that nicotine elicits changes in neural cell replication and differentiation, compromising neuronal maturation, leading to long-lasting alterations in the structure of key brain regions involved in cognition, learning, and memory possibly leading to deficits in synaptic neurochemistry and behavioral performance.

9.2.2 Impact on Hormone Synthesis Relevant for Brain Development

Nicotine causes changes in the synthesis of both adrenal and gonadal hormones, both of which may disrupt normal brain development. When timed-pregnant rats were treated with nicotine delivered by an osmotic minipump for either 1 week (2 or 6 mg/kg/d from gestation day 12) or 2 weeks (6 mg/kg/d from gestation day 8), their male offspring showed a decrease in activity of aromatase (enzyme that converts androgens to estrogens and is known to be important in sexual brain differentiation) to female levels at postnatal day 6. In the same study, low doses of prenatal nicotine (2 mg/kg/day) are sufficient to cause increases in plasma corticosterone in G18 male fetuses.[377] Nicotine has been reported to inhibit aromatase activity in the human placenta cell line, which may also interfere with brain sexual differentiation in the male.[378] In a 90-day inhalation study on rats, nicotine induced a substantial upregulation on liver gene expression of 17α-hydroxylase-17,20-lyase (CYP17A); a member of an enzyme complex essential for production of adrenal glucocorticoids and androgens, as well as gonadal androgens, has been observed (manuscript submitted). This effect may explain gender-related responses in several studies on rodents. Nicotine administered by injection from gestational days 1–20 improved learning in a two-way active avoidance task in females but impaired the process in male rats.[379] Similarly, adult male rats exposed to nicotine via injection on postnatal days 8–12 were impaired in learning an auditory-cued active avoidance task,[380] while no significant effect on learning was observed in a visual and auditory-cued active avoidance task in female rats exposed to nicotine prenatally.[381] It is especially interesting that males appear to be more susceptible to those challenges because boys are more likely to be diagnosed with ADHD and may display a different spectrum of symptoms compared with girls.[382] Enhanced learning in a trace fear conditioning task was observed in male offspring of C57BL6/J mice administered with nicotine in drinking water, and no significant alteration in learning was observed in females.[213]

9.2.3 Impact of Developmental Exposure to Nicotine on nAChRs Expression

Gestational nicotine exposure changes the density of nAChR expression, during both the fetal period and later postnatal life. In the study mentioned above, Slotkin et al. investigated the fetal and postnatal development of binding sites for radioactively labeled nicotine in brain regions of rats whose mothers received nicotine injections or infusions. Fetal exposure to nicotine produced an elevation in binding detectable during the course of drug exposure (gestational day 18), a finding similar to nicotine's effects in mature brain. However, examination of the subsequent developmental pattern of [3H]nicotine binding indicated a generalized disruption of receptor acquisition with alterations persisting well after the discontinuation of drug exposure. The greatest effect was seen the cerebellum, which is a region with relatively few receptor sites.[383] Similar results were reported by Hagino and Lee.[384]

Hyperactivity in male rat offspring induced by prenatal nicotine exposure (6 mg/kg/day of nicotine via osmotic pump on gestation day 4) appears to be associated with an increase in neuronal nicotinic receptors in the cortex and possibly the striatum. An increase in nicotinic receptors density, without change in binding affinity, is determined by [3H]cytisine binding in tissue homogenate. In this study, hyperactivity was detected in control-treated males as well but without an increase in cortical nicotinic receptors. This suggests that cortical or striatal nicotinic receptors may not be critical in development of this behavioral disorder, and other mechanisms could be involved as well.[385]

Nicotine exposure during critical brain development periods induces modifications of specific nAChRs. Neonatal nicotine administration (6 mg/kg/day from postnatal days 1–8) induced significant upregulation of heteromeric but not homomeric nAChRs in different brain regions: the hippocampus, cortex, and thalamus, evaluated by autoradiography using (125I)-epibatidine and (125I)-alpha-bungarotoxin. Changes in subunit mRNA expression were not observed by *in situ* hybridization, suggesting that the changes occurred posttranscriptionally. No effect of chronic nicotine on receptor expression was detected in the medial habenula, suggesting that heteromeric nAChRs $\alpha4\beta2$ are more sensitive to upregulation than $\alpha7$ nAChRs, which is consistent with the subtype selective upregulation observed in adulthood.[386] More recently, Chen K et al. exposed early postnatal rats to nicotine through maternal milk and examined the changes in the function of $\alpha2^*$-nAChR-containing oriens-lacunosum moleculare (OLM) cells during adolescence in rats. OLM cells expressing $\alpha2^*$ nAChR are an important component of hippocampal circuitry, gating information flow, and long-term potentiation (LTP) in the CA1 region. Results revealed significant decrease not only in the number of $\alpha2$-mRNA-expressing interneurons in the stratum oriens/alveus but also $\alpha2^*$-nAChR-mediated responses in OLM cells suggesting this region in the brain to be possibly an important target of further studies for identifying the mechanisms underlying the cognitive impairment induced by maternal smoking during pregnancy.[387]

It should be noted that, while nicotine-induced nAChR regulation during important phases of brain development may contribute significantly to neurological consequences of developmental nicotine exposure, other mechanisms are potentially more critical, as suggested by observations of relatively poor correlation between nAChR regulation and behavioral alterations.[388]

9.2.4 Impact of Prenatally Administered Nicotine on Neurotransmitter Systems

Prenatal exposure to nicotine has an impact on neurotransmitters in the developing brain. Alterations in the dopaminergic and adrenergic systems as a result of developmental nicotine exposure were reported in rats.[389] This is important because catecholaminergic signaling in the prefrontal cortex is thought to be disrupted in ADHD patients.[390] Other brain structures can be affected as well. Dopamine concentration was decreased in the ventral tegmental area and striatum but was increased in the substantia nigra of the hyperactive offspring of gestationally exposed female rat.[206] Nicotine exposure can increase adrenergic turnover in the forebrain during gestation and is followed by decreased turnover postnatally, an effect that persists into adulthood in males.[391] Prenatally administered nicotine-induced alterations in the serotonergic system of developing brain that may contribute to some aspects of ADHD development have been reported as well. Gestational administration of nicotine (6 mg/kg/day via subcutaneous injections or infusion pumps) in rats increased serotonin transporter expression in the brain.[389] Aberrant serotonergic system has been reported to have some implications on ADHD development.[392]

9.2.5 Summary

Prenatal nicotine exposure via maternal smoking has been shown to have concerning effects on human brain development. Several animal studies have indicated that nicotine exposure during development could affect a wide variety of biological processes at the molecular, cellular, tissue, and functional levels compromising neuronal maturation, leading to long-lasting alterations in the structure of key brain regions involved in cognition, learning, and memory.

nAChRs in the brain are expressed at an early stage of development, and there is considerable evidence that they play a regulatory role. Neurodevelopmental events known to be regulated by ACh (the endogenous nAChR agonist) can be affected by the exposure to nicotine. As a consequence, aberrant neuronal structures could result in altered processing of sensory input in individuals exposed to tobacco during development and may lead to disorders such as ADHD.

9.3 nAChR AGONISTS

Recent literature shows a shift in the way researchers consider development and design of drugs to treat diseases with complex etiological pathways (i.e., diseases with multiple drug targets), such as psychiatric and neurodegenerative diseases.[393–395] A drug with a single-target mechanism of action cannot always compensate for or correct a complex pathway. This suggests that a complex pathway of disease should be treated either with a combination of molecules, each acting on different disease-relevant pathways (polypharmacy), or with one molecule that possesses promiscuous activity and acts on different pathways (multiple mechanism drugs). Because of the complex molecular actions of nAChRs that lead to possible neuroprotection, and given the lack of efficient therapies, companies have invested substantial resources in developing various nAChR ligands aimed at offering patients suffering from neurodegenerative and psychiatric illnesses an improved quality of life. A number of the molecules developed were inspired by the chemical structure of nicotine, while taking into account that nicotine has a strong addictive potential (a selection of compounds are shown in Table 1). Most of the developed compounds (including full and partial agonists and positive allosteric modulators) are selective for receptors composed of the canonical subunits, α7 and α4β2. Far less research has been conducted on the other receptor subunits.

TABLE 1 Selected Nicotinic Acetylcholine Receptor (nAChR[a]) Agonists Developed for Psychiatric and Neurodegenerative Diseases

Compound	Mode of Action	Indication	Development Phase	Reference or Trial Number
Varenicline	α4β2 partial agonist	Smoking withdrawal	Marketed	1,2
		AD[b]	Phase 2	3
Lobeline	α4β2 agonist	ADHD[c]	Phase 2	NCT00664703[d] 4
Altinicline/SIB1508Y	α4β2 agonist	PD[e]	Phase 2	5
Pozanicline/ABT-089	α4β2 agonist	AD	Phase 2	NCT00069849 NCT00809510
Pozanicline/ABT-089	α4β2 agonist	AD	Phase 2	NCT00069849 NCT00809510
		ADHD	Phase 2	NCT00686933
AZD1446/TC-6683	α4β2 agonist	AD	Phase 2	NCT01039701 NCT01125683
		ADHD	Phase 2	NCT01012375 6
Ispronicline/AZD3480	α4β2 agonist	AD	Phase 2	NCT01466088
		Cognition disorders	Phase 2	NCT00109564
		ADHD	Phase 2	NCT00683462
Sofinicline/ABT-894	α4β2 agonist	ADHD	Phase 2	NCT00429091
AQW051	α7 agonist	Cognition disorders	Phase 2	NCT01730768

Continued

TABLE 1 Selected Nicotinic Acetylcholine Receptor (nAChR[a]) Agonists Developed for Psychiatric and Neurodegenerative Diseases—cont'd

Compound	Mode of Action	Indication	Development Phase	Reference or Trial Number
		Drug-induced dyskinesia	Phase 2	NCT01474421
		AD/MCI[f]	Phase 2	NCT00582855
Bradanicline/TC-5619	α7 agonist	Cognition disorders (adjunctive treatment for schizophrenic patients)	Phase 2	NCT01488929 7
		AD	Phase 1	NCT01254448
		ADHD	Phase 2	NCT01124708 NCT01472991
Nelonicline/ABT-126	α7 agonist	Cognition disorders	Phase 2	NCT01834638 NCT01655680
		AD	Phase 2	NCT00948909 NCT01527916
SSR180711	α7 agonist	AD	Phase 2	NCT00602680
RG3487/MEM3454	α7 agonist/5-HT3 receptor antagonist	AD	Phase 2	NCT00454870 NCT00884507
		Schizophrenia	Phase 2	NCT00604760 8
GTS-21	α7 partial agonist	AD	Phase 2	NCT00414622
		Cognition disorders	Phase 2	NCT01400477

		ADHD	Phase 1/2	NCT00419445
		Schizophrenia	Phase 2	NCT00100165 9
Encenicline/EVP-6124	α7 partial agonist	AD	Phase 3	NCT01969136
		Cognition disorders	Phase 3	NCT01714713 NCT01714661 NCT01716975
ABT-418	Nonselective nicotinic receptor agonist	AD	Phase 2	10
		ADHD	Pilot trial	11

[a] *nAChR, nicotinic acetylcholine receptor.*
[b] *AD, Alzheimer's disease.*
[c] *ADHD, attention deficiency hyperactivity disorder.*
[d] *NCT, National Clinical Trial.*
[e] *PD, Parkinson's disease.*
[f] *MCI, mild cognitive impairment.*

1. Tonstad S, Tonnesen P, Hajek P, et al. Effect of maintenance therapy with varenicline on smoking cessation: a randomized controlled trial. *JAMA* 2006;**296**:64–71.
2. Gonzales D, Rennard SI, Nides M, et al. Varenicline, an alpha4beta2 nicotinic acetylcholine receptor partial agonist, vs sustained-release bupropion and placebo for smoking cessation: a randomized controlled trial. *JAMA* 2006;**296**:47–55.
3. Kim SY, Choi SH, Rollema H, et al. Phase II crossover trial of varenicline in mild-to-moderate Alzheimer's disease. *Dement Geriatr Cogn Disord* 2014;**37**:232–45.
4. Martin CA, Nuzzo PA, Ranseen JD, et al. Lobeline effects on cognitive performance in adult ADHD. *J Atten Disord* 2013. doi:10.1177/1087054713497791.
5. Parkinson Study Group. Randomized placebo-controlled study of the nicotinic agonist SIB-1508Y in Parkinson disease. *Neurology* 2006;**66**:408–10.
6. Jucaite A, Ohd J, Potter AS, et al. A randomized, double-blind, placebo-controlled crossover study of alpha4beta 2* nicotinic acetylcholine receptor agonist AZD1446 (TC-6683) in adults with attention-deficit/hyperactivity disorder. *Psychopharmacology (Berl)* 2014;**231**:1251–65.
7. Lieberman JA, Dunbar G, Segreti AC, et al. A randomized exploratory trial of an alpha-7 nicotinic receptor agonist (TC-5619) for cognitive enhancement in schizophrenia. *Neuropsychopharmacology* 2013;**38**:968–75.
8. Umbricht D, Keefe RS, Murray S, et al. A randomized, placebo-controlled study investigating the nicotinic alpha7 agonist, RG3487, for cognitive deficits in schizophrenia. *Neuropsychopharmacology* 2014;**39**:1568–77.
9. Tregellas JR, Tanabe J, Rojas DC, et al. Effects of an alpha 7-nicotinic agonist on default network activity in schizophrenia. *Biol Psychiatry* 2011;**69**:7–11.
10. Potter A, Corwin J, Lang J, Piasecki M, Lenox R, Newhouse PA. Acute effects of the selective cholinergic channel activator (nicotinic agonist) ABT-418 in Alzheimer's disease. *Psychopharmacology (Berl)* 1999;**142**:334–42.
11. Wilens TE, Biederman J, Spencer TJ, et al. A pilot controlled clinical trial of ABT-418, a cholinergic agonist, in the treatment of adults with attention deficit hyperactivity disorder. *Am J Psychiatry* 1999;**156**:1931–7.

Chapter 10

Other Compounds From Tobacco With Potential Impact on Neurodegenerative Diseases

In the context of its pharmacological actions, nicotine is the most studied tobacco smoke constituent. Nonetheless, nicotine is not the only constituent of tobacco leaf and cigarette smoke that displays pharmacological activity. Given the complexity of data from clinical studies on nicotine and epidemiological data on smoking and AD and PD (see Chapters 1.3.1, 1.3.3, 2.3.1, and 2.3.3), it is possible that nicotine exerts its effects together with other bioactive, possibly inhalable, smoke constituents that should be studied separately, that is, when decoupled from tobacco smoke and overall harmful effects of smoking. This section describes some tobacco/smoke compounds for which neurological effects have been reported.

10.1 MONOAMINE OXIDASE INHIBITORS FOR NEURODEGENERATIVE DISEASE THERAPY

Monoamine oxidase (MAO) A and B are mitochondrial-bound isoenzymes that catalyze oxidative deamination of dietary amines and monoamine neurotransmitters, such as serotonin, norepinephrine, dopamine, β-phenylethylamine, and other trace amines (Fig. 1). Type A and type B MAOs are designated according to their substrate specificity and sensitivity to distinct inhibitors. MAO A preferentially oxidizes serotonin and norepinephrine, while MAO B oxidizes β-phenylethylamine and benzylamine. Dopamine is oxidized by both isoenzymes, which are coded by different genes closely localized on the X chromosome.[396]

MAO function is an important element in the etiology of age-regulated neurodegenerative disorders such as PD, AD, and depression.[397,398] It is well established that MAO activity is decreased in smokers[399–402]; whole-body positron emission tomography (PET)-scanning studies performed by Fowler et al. clearly show decreased MAO activity in the brain[403–405] and peripheral tissues[406–408] of tobacco smokers.

Fowler et al. observed increased MAO B expression with age in the brains of normal healthy human subjects, measured using [^{11}C]L-deprenyl-D2 and

Nicotine and Other Tobacco Compounds in Neurodegenerative and Psychiatric Diseases.

https://doi.org/10.1016/B978-0-12-812922-7.00010-X

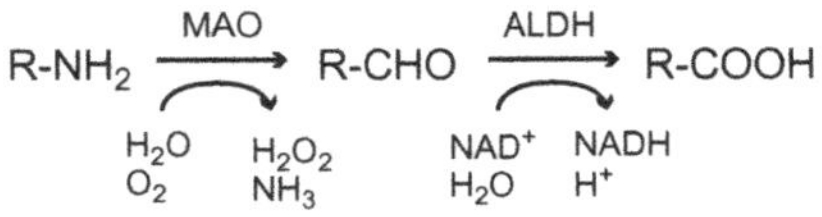

$$R\text{-}NH_2 \xrightarrow[H_2O,\ O_2 \;\to\; H_2O_2,\ NH_3]{MAO} R\text{-}CHO \xrightarrow[NAD^+,\ H_2O \;\to\; NADH,\ H^+]{ALDH} R\text{-}COOH$$

FIG. 1 Monoamine oxidase (MAO) catalyzes oxidative deamination of monoamines. Monoamines are degraded by MAO to their correspondent aldehydes (R-CHO). This reaction also produces ammonia (NH_3) and hydrogen peroxide (H_2O_2). These aldehydes are further oxidized by aldehyde dehydrogenase (ALDH) into carboxylic acids (R-COOH). Reduced form of nicotinamide adenine dinucleotide (NADH) is a critical cofactor for this latter reaction.

PET.[409] Similarly, MAO B activity was increased in platelets from AD and PD patients.[410,411] Activity and mRNA levels of MAO A and B are increased in the frontal cortex of AD brains,[412,413] while increased MAO A activity reduces the levels of noradrenaline, serotonin, and melatonin, the reduction of which may disturb sleep-wake rhythms and induce cognitive decline and depression in AD patients.[414]

The impact of smoking on MAO activity has long attracted the attention of scientists. An early study conducted by Yu and Boulton reported that phosphate-buffered cigarette smoke extract or tobacco extract inhibited mitochondrial MAO (the isoenzymes were identified and characterized later) isolated from the lung tissue of Wistar rats. The authors exposed rat lung tissue to either cigarette smoke solution or smoke directly and then isolated mitochondrial MAO and tested its activity. A drastic reduction in enzyme activity was detected with exposure to any form of smoke. The same study reported that saliva from human smokers inhibits MAO activity.[415] Comparably, an earlier study found that cigarette smoke exposure inhibits MAO activity in mouse skin. Skin homogenates of exposed mice were incubated with ^{14}C-labeled serotonin and tyramine, and the quantity of metabolites from these substrates is determined.[416] Hexane extracts from tobacco leaf and smoke reduced mitochondrial MAO B activity from baboon liver extracts.[417] A recent study reported that cigarette smoke extract, created by bubbling smoke from commercial cigarettes through sterile saline, exhibited potent MAO A and B inhibition in rat brain mitochondrial homogenates.[418]

In vitro studies have further revealed details on tobacco-related MAO inhibitors. Harman is a specific competitive MAO A inhibitor ($IC_{50}=0.34\,\mu M$), while norharman is a nonspecific competitive MAO A and B inhibitor ($IC_{50}=6.5$ and $4.7\,\mu M$, respectively). Both compounds readily cross the blood-brain barrier (BBB) and accumulate in the brain, contributing to the inhibitory effect of tobacco smoke on MAOs.[419] A benzoquinone, 2,3,6-trimethyl-1,4-naphthoquinone, with weak MAO-inhibiting activity (K_i, 3–6 μM),[420] and 2-naphthylamine, with a 10-fold lower potency, have also been isolated from tobacco leaves.[421] Mendez-Alvarez et al. showed that MAO inhibitory activity of tobacco smoke may be derived not only from the tobacco smoke components themselves but also from the products of the reactions of cigarette smoke components with endogenous or exogenous compounds, resulting in formation of bioactive compounds.[422] Finally, nitric oxide (NO), a major component of cigarette smoke (reaching ppm levels in smoke), displays MAO-inhibiting activity, as indicated by the observation that the NO donor S-nitroso-N-acetylpenicillamine (SNAP) effectively inhibits MAO activity in the range of 0.4–40 μM.[423] Table 1 summarizes MAO inhibitors from tobacco plants or smoke.

MAO inhibitors have a long history of use as medications prescribed for the treatment of depression. They are particularly effective in treating atypical depression.[424] However, because of potentially lethal dietary and drug interactions (an extreme hypertension response caused by noradrenaline release from synaptic vesicles by tyramine, a by-product of MAO A metabolism in the gut wall, and found in cheese and other foods), newer MAO inhibitors, reversible and with higher selectivity, have been developed.[425] Selegiline was the first established, irreversible MAO B selective inhibitor without the "cheese effect" and therefore one of the earliest to be administered as an antidepressant.[426,427]

In an open, long-term (9 years) clinical study, Birkmayer et al. observed possible neuroprotective activity of selegiline when used in combination with L-DOPA.[428] Inhibition of MAO activity is not the only mechanism through which MAO inhibitors exert their neuroprotective features. For example, selegiline displays complex pharmacological activities, including reduction of reactive oxygen species (ROS) generation by inhibition of MAO-dependent oxidation, direct scavenging of hydroxyl radicals, and increased activity of antioxidant enzymes such as superoxide dismutase (SOD) 1 and 2, catalase, and thioredoxin.[429–433] Selegiline and its metabolites inhibit reuptake of dopamine, increase dopamine release, and inhibit presynaptic catecholamine receptors. Interestingly, chronic administration of selegiline at low doses has been shown to increase lifespan in rats and PD patients.[434–436] Rasagiline is another MAO B inhibitor with higher affinity to MAO B and more potent inhibitory activity than selegiline both *in vivo* and *in vitro*.[437] It is often used in combination with L-DOPA in the treatment of parkinsonian patients.

Neuroprotective features of selegiline and rasagiline have been reported in several animal models in which dopaminergic,[438–440] noradrenergic,[441,442] and cholinergic neurotoxins,[443] excitotoxins, ischemia, and other insults were tested.[444–447] Similar findings for a protective role of MAO B inhibitors were reported using cellular models of apoptosis induced by neurotoxins, oxidative stress, and serum withdrawal.[448–454] Mechanistically, the neuroprotective action of MAO inhibitors is likely exerted through their intervention in stepwise activation of apoptotic signaling. MAO inhibitors prevent the decline in membrane potential and cytochrome c release in isolated mitochondria, stabilizing the mitochondrial intermembrane space.[454,455] Calcium efflux is the initial signal in the mitochondrial apoptotic cascade. Wu and colleagues reported that rasagiline and selegiline inhibit mitochondrial calcium ion efflux through the mitochondrial permeability transition pore in a dose-dependent manner in

TABLE 1 Monoamine Oxidase (MAO) Inhibitors From Tobacco Plants or Smoke

Compound Name	Reference
2,3,6-Trimethyl-1,4-naphthoquinone	1
Farnesylacetone	2
2-Naphthylamine	3
Harman	4
Norharman	4
trans,trans-Farnesol	5
Acetaldehyde	6
Salsolinol	7
Nornicotine	8, 9
Anatabine	9
Anabasine	9
Nitric monoxide (NO)	10
2-Naphthylamine	11

1. Khalil AA, Steyn S, Castagnoli N. Isolation and characterization of a monoamine oxidase inhibitor from tobacco leaves. *Chem Res Toxicol* 2000;**13**:31–5.
2. Castagnoli K, Steyn SJ, Magnin G, et al. Studies on the interactions of tobacco leaf and tobacco smoke constituents and monoamine oxidase. *Neurotox Res* 2002;**4**:151–60.
3. Hauptmann N, Shih JC. 2-Naphthylamine, a compound found in cigarette smoke, decreases both monoamine oxidase A and B catalytic activity. Life Sci 2001;**68**:1231–41.
4. Herraiz T, Chaparro C. Human monoamine oxidase is inhibited by tobacco smoke: β-carboline alkaloids act as potent and reversible inhibitors. *Biochem Biophys Res Commun* 2005;**326**: 378–86.
5. Khalil AA, Davies B, Castagnoli N. Isolation and characterization of a monoamine oxidase B selective inhibitor from tobacco smoke. *Bioorg Med Chem* 2006;**14**:3392–8.
6. Talhout R, Opperhuizen A, van Amsterdam JG. Role of acetaldehyde in tobacco smoke addiction. *Eur Neuropsychopharmacol* 2007;**17**:627–36.
7. Naoi M, Maruyama W, Nagy GM. Dopamine-derived salsolinol derivatives as endogenous monoamine oxidase inhibitors: occurrence, metabolism and function in human brains. *Neurotoxicology* 2004;**25**:193–204.
8. Clemens KJ, Caillé S, Stinus L, Cador M. The addition of five minor tobacco alkaloids increases nicotine-induced hyperactivity, sensitization and intravenous self-administration in rats. *Int J Neuropsychopharmacol* 2009;**12**:1355–66.
9. Williams JR, Delorenzo RJ, Burton HR. Monoamine oxidase (MAO) inhibitors and uses thereof. Google Patents; 2003.
10. Muriel P, Pérez-Rojas JM. Nitric oxide inhibits mitochondrial monoamine oxidase activity and decreases outer mitochondria membrane fluidity. *Comp Biochem Physiol C: Toxicol Pharmacol* 2003;**136**:191–7.
11. Hauptmann N, Shih JC. 2-Naphthylamine, a compound found in cigarette smoke, decreases both monoamine oxidase A and B catalytic activity. *Life Sci* 2001;**68**:1232–1241.

SH-SY5Y cells.[456] Induction of prosurvival, antiapoptotic genes, such as Bcl-2 and different neurotrophic factors, contributes to the neuroprotective activity of propargylamine derivatives, as thoroughly reviewed by Naoi and Maruyama.[457]

Rasagiline has been used to design analogs by addition of pharmacophores that act on other neurological targets. This multitarget approach may prove successful in the development of new and more effective therapies targeting multifaceted neurodegenerative diseases. Table 2 summarizes the effects of rasagiline and structurally related compounds on amyloid beta (Aβ) accumulation, along with their putative mechanisms. Most of these drugs have serious shortcomings (e.g., neurotoxicity or inability to cross the BBB), which have prevented them from reaching clinical settings.

Selegiline is mainly used in the treatment of PD and can be used on its own or in combination with another agent, most often L-DOPA.[397] For newly diagnosed PD patients, selegiline has been reported to slow disease progression, although this has not been widely accepted, and the methodology has been rejected by the Food and Drug Administration (FDA). Rasagiline is used as a monotherapy in early PD or as an adjunct therapy in more advanced cases.[458]

TABLE 2 Effects of Rasagiline-Related Propargylamine Derivatives on Aβ[a] Accumulation

Compounds Structurally Related to Rasagiline	Effect on Aβ Accumulation	Model System	Reference
TV3279[b], TV3326, rasagiline, *N*-propargylamine	Increased sAPPα[c] Prevented increase in holo-APP protein	SH-SY5Y, PC12, SK-N-SH cells	1
Selegiline	Increased sAPPα secretion	SK-N-SH, PC12 cells	2
M-30	Reduced Aβ secretion and cellular β-CTF[d] levels Increased sAPPα secretion	CHO[e] cells transfected with APP "Swedish" mutation	3

[a] *Aβ, amyloid beta.*
[b] *TV3279, S-isomer of TV3326.*
[c] *sAPPα, soluble extracellular N-terminal fragment of amyloid precursor protein (APP).*
[d] *β-CTF, beta C-terminal fragment.*
[e] *CHO, Chinese hamster ovary cells.*

1. Bar-Am O, Weinreb O, Amit T, Youdim MB. Regulation of Bcl-2 family proteins, neurotrophic factors, and APP processing in the neurorescue activity of propargylamine. *FASEB J* 2005;**19**:1899–901.
2. Yogev-Falach M, Bar-Am O, Amit T, Weinreb O, Youdim MB. A multifunctional, neuroprotective drug, ladostigil (TV3326), regulates holo-APP translation and processing. *FASEB J* 2006;**20**: 2177–9.
3. Avramovich-Tirosh Y, Amit T, Bar-Am O, Zheng H, Fridkin M, Youdim MB. Therapeutic targets and potential of the novel brain- permeable multifunctional iron chelator-monoamine oxidase inhibitor drug, M-30, for the treatment of Alzheimer's disease. *J Neurochem* 2007;**100**:490–502.

Clinical data suggest a neuroprotective effect, but are not conclusive. The FDA has repeatedly denied Teva Pharmaceuticals' request for an on-label indication for neuroprotection in PD. Similar to PD, MAO inhibitors are used in combination with other therapeutic agents in AD with limited success.

10.2 OTHER TOBACCO COMPOUNDS AND NEUROPROTECTION

10.2.1 Anatabine and Anabasine

Anabasine and anatabine are tobacco alkaloids structurally similar to nicotine and are ligands for nAChRs.[459–461]

A recent study examined the effect of anatabine (Fig. 2) on Aβ production in a model of AD. Paris et al. showed that anatabine lowers Aβ production mainly by disrupting β-cleavage of APP and lowering NF-κB activation and β-secretase 1 expression (the rate-limiting enzyme for Aβ production) *in vitro* using a cell line overexpressing human APP. Acute anatabine treatment for four days and chronic treatment for 6.5 months significantly lowered brain soluble Aβ1–40 and Aβ1–42 levels in a transgenic mouse model of AD.[462,463] In 2013, the same group published a study on the effects of anatabine on tau phosphorylation and oligomerization in a pure tauopathy model of AD (Tg Tau P301S). Anatabine reduced motor impairments and tau phosphorylation and oligomerization in the brain and spinal cord of Tg Tau P301S mice, suggesting that anatabine should be further explored as a treatment for tauopathies and AD in particular. Based on previous work from this group demonstrating the antiinflammatory effects of anatabine and the structural similarity of nicotine and anatabine, Paris et al. proposed that anatabine acts via nAChRs on neuronal cell membranes. This leads to the activation of the prosurvival PI3K/Akt pathway that inhibits GSK-3β, a critical component for Aβ accumulation and tau hyperphosphorylation, which is followed by tau oligomerization and neurodegeneration.[464] Less is known about the nicotinic receptor binding of anatabine, but it has been identified as a ligand of the α3β4 subtype.[460]

Paris et al. also reported that orally administered anatabine markedly suppressed neurological deficits associated with development of EAE, an experimental mouse model of MS. Analysis of peripheral cytokine production revealed that anatabine significantly reduced Th1 and Th17 cytokines, which are known to contribute to EAE development. Furthermore, anatabine appears to significantly suppress STAT3 and p65 NF-kB phosphorylation in the spleen

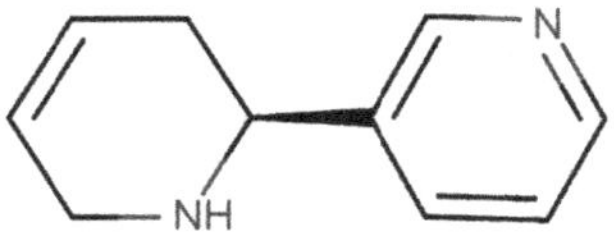

N
NH

FIG. 2 Chemical structure of anatabine.

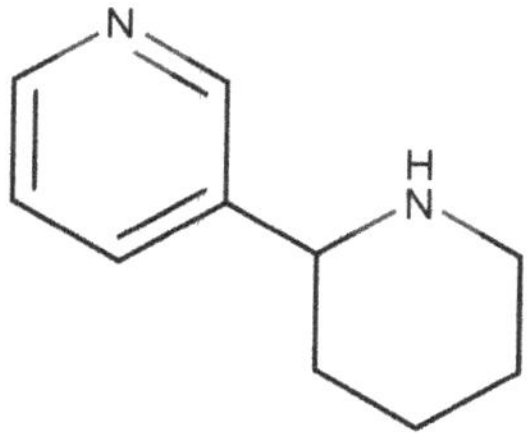

FIG. 3 Chemical structure of anabasine.

and brain of EAE mice. These two transcription factors regulate a large array of inflammatory genes, including cytokines, suggesting a mechanism by which anatabine antagonizes proinflammatory cytokine production. Additionally, anatabine alleviates macrophage/microglia infiltration and astrogliosis and significantly prevents demyelination in the spinal cord of EAE mice.[465]

Anabasine (Fig. 3) is a partial agonist with lower affinity than nicotine for $\alpha4\beta2$ nAChRs, but greater affinity than nicotine for the $\alpha7$ subtype, for which it is a full agonist and hypothesized to exert most of its *in vivo* effects. GTS-21, a derivative of the structurally similar anabaseine,[a] is one of the first compounds created showing selective agonist properties for the $\alpha7$ receptor.[466] Different effects were seen with anabasine and anatabine on memory and attention in adult female Sprague-Dawley rats, trained on a win-shift spatial working and reference memory task in a 16-arm radial maze, or a visual signal detection operant task to test attention. Anabasine, but not anatabine, significantly reversed dizocilpine (MK-801)-induced memory impairment. With dizocilpine-induced signal detection attentional impairment, a reversal was also observed, but this time, anatabine, but not anabasine, was responsible for this reversal. It is likely that both compounds exert different nAChR effects that may explain the observed differential cognitive effects.[467]

In addition to the effects of anatabine and anabaseine, Dwoskin and colleagues have demonstrated concentration-dependent stimulation of ^{3}H-labeled dopamine release in rat striatal slices using two other minor tobacco alkaloids, norcotinine and methylanabasine.[468]

10.2.2 β-Carboline

Two β-carboline alkaloids, norharmane (Fig. 4) and harmane (1-methyl-β-carboline), were identified by gas chromatography-mass spectrometry (GC-MS), quantified, and isolated from mainstream smoke by solid-phase extraction and high-performance liquid chromatography (HPLC). These compounds are competitive, reversible, and potent inhibitors of MAO enzymes.[419] A β-carboline scaffold has been utilized and targeted in traditional Chinese medicine. An *in vitro* antiinflammatory effect of β-carboline was assessed in the RAW

a. Produced by *Nemertines* and *Aphaenogaster* ants.

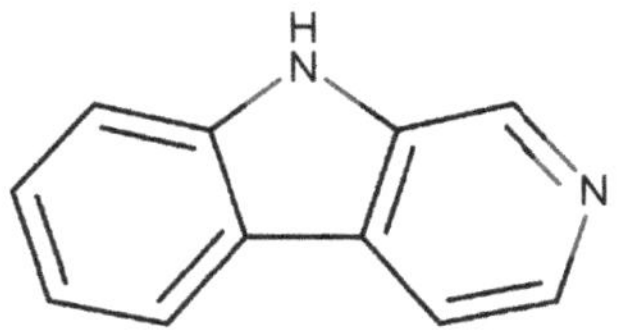

FIG. 4 Chemical structure of the β-carboline compound norharmane.

264.7 macrophage cell line.[469] Following LPS treatment, three structurally related β-carboline compounds suppressed proinflammatory cytokine expression, specifically TNF-α and interleukin (IL) 6. Although all compounds displayed antiinflammatory effects in a dose-dependent manner, there were differences in efficacy (IC_{50} ranging from 4 to almost 10 μM). A comprehensive structure-activity relationship analysis, using TNF-α and IL-6 expression in cell-based assays, confirmed the antiinflammatory potential of this class of compounds.[470] Rats treated with 2 μmol 9-methyl-β-carboline for 10 days showed improved spatial learning in the radial maze, elevated dopamine levels in the hippocampal formation, and more complex and elongated dendritic trees and higher spine numbers on granule neurons of the dentate gyrus. This suggests that 9-methyl-β-carboline acts as a cognitive enhancer in a hippocampus-dependent task and that the behavioral effects may be associated with a stimulatory impact on hippocampal dopamine levels and dendritic and synaptic proliferation.[471]

10.2.3 β-Sitosterol

β-Sitosterol is a plant sterol with a chemical structure similar to that of cholesterol (Fig. 5), which is present in tobacco and cigarette smoke.[472]

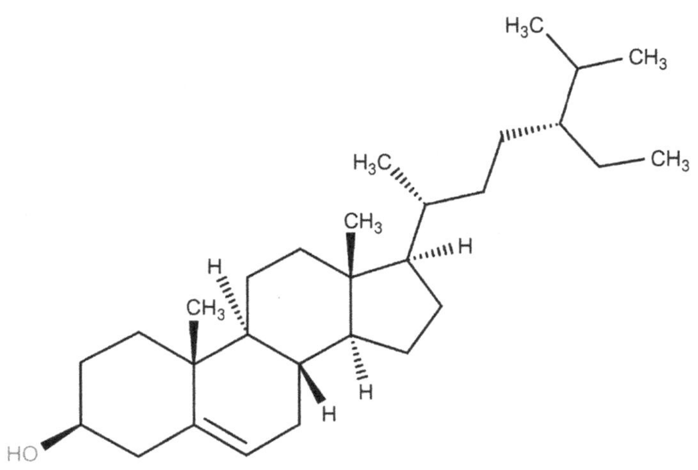

FIG. 5 Chemical structure of β-sitosterol.

Direct evidence for a neuroprotective effect of β-sitosterol has not been found. A possible indirect link, though the evidence is inconclusive, was provided by Shi et al., showing that β-sitosterol incorporation into the mitochondrial membrane enhances mitochondrial function by promoting inner mitochondrial membrane fluidity.[473]

10.2.4 Caffeic Acid

Caffeic acid (Fig. 6) is one of the chemicals isolated from the bark of *Eucalyptus globulus*.[474] It is also found in the freshwater fern *Salvinia molesta*,[475] in the mushroom *Phellinus linteus*,[476] and in tobacco.[477]

Caffeic acid added to food for 10 weeks attenuated colonic inflammation in the dextran sulfate sodium (DSS)-induced mouse model of colitis.[478] Further, hydrocaffeic acid, a metabolite of caffeic acid, reduced the level of tissue damage caused by DSS when orally administered at a dose of 50 mg/kg. Cytokine perturbations showed a reduction trend in expression of IL-1β, IL-8, and TNF-α, malondialdehyde (MDA) levels, and oxidative DNA damage in colon mucosa.[479] Improvements in memory deficits and cognitive impairments attributed to caffeic acid (oral dose of 10 or 50 mg/kg/day for 2 weeks) were observed in an $A\beta_{25-35}$-injected AD mouse model.[480,481]

10.2.5 Cembranoids

Cembranoids are 14-carbon cembrane ring cyclic diterpenoids found in tobacco leaves and flowers, marine invertebrates (soft coral gorgonian species), insects, and even vertebrates.[482] The most abundant tobacco cembranoids are (1S,2E,4R,6R,7E,11E)-cembra-2,7,11-triene-4,6-diol (also termed 4R- cembratriene) and its diastereoisomer (1S,2E,4S,6R,7E,11E)-cembra-2,7,11-triene-4,6-diol (also termed 4S-cembratriene). 4R-cembratriene protects against acute neurotoxicity induced by *N*-methyl-D-aspartate (NMDA) and the organophosphorus compound paraoxon in hippocampal slices, acting via α7 and α4β2 nAChRs.[483–485] In addition, specific 4R-cembratriene analogs protect

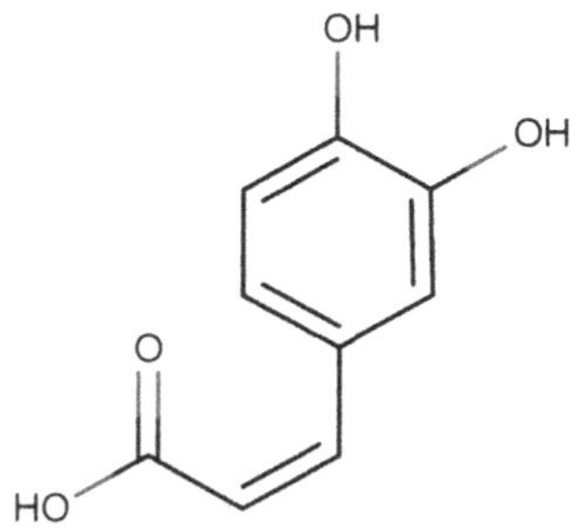

FIG. 6 Chemical structure of caffeic acid.

hippocampal slices against neurotoxicity of diisopropylfluorophosphate, a surrogate of sarin, the chemical war nerve agent.[486,487] The neuroprotective activity of 4R is mediated via activation of the PI3K/AKT antiapoptotic cascade. Consequently, the proapoptotic enzyme, GSK-3β, is inactivated, leading to a reduction in NMDA-induced apoptosis. The RAF/MEK/ERK cascade is apparently not involved in 4R-mediated neuroprotection.[483]

Tobacco cembranoids also display antiinflammatory properties that may enhance their neuroprotective properties. The 4S and 4R cembranoids inhibit prostaglandin synthesis, with lower IC_{50} values than acetylsalicylic acid.[488]

10.2.6 Chlorogenic Acid

Chlorogenic acid is a compound found in a wide variety of foods and beverages, including fruits, vegetables, olive oil, spices, wine, and coffee.[489] It is also found in tobacco leaves.[490] Chlorogenic acid (Fig. 7) has been extensively investigated in neurodegenerative diseases because of its antiinflammatory activity, which is attributed to microglia activation,[491] and antioxidant brain activity.[492]

A study by Kim et al. reported that chlorogenic acid suppressed apoptotic nuclear condensation induced by hydrogen peroxide in neuronal cells (12.5–100 μM for an hour). The findings indicate that treatment with chlorogenic acid could protect against neurological degeneration associated with oxidative stress in the brain.[493] Chlorogenic acid significantly improved the impairment of short-term or working memory induced by scopolamine and significantly reversed cognitive impairments and decreased escape latencies in mice. *ex vivo*, chlorogenic acid was shown to inhibit AChE activity in the hippocampus and frontal cortex, indicating that it may exert antiamnesic activity.[494]

10.2.7 Cotinine

An increasing body of *in vivo* and *in vitro* evidence suggests that cotinine (Fig. 8), the predominant nicotine metabolite in mammalian species, may also

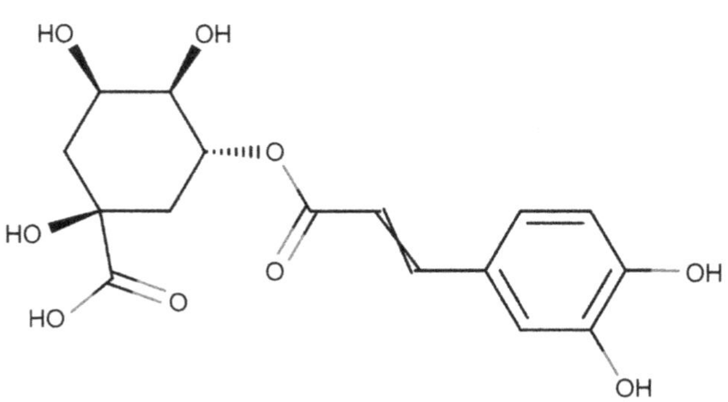

HO OH
O
HO
O OH
HO O
OH

FIG. 7 Chemical structure of chlorogenic acid.

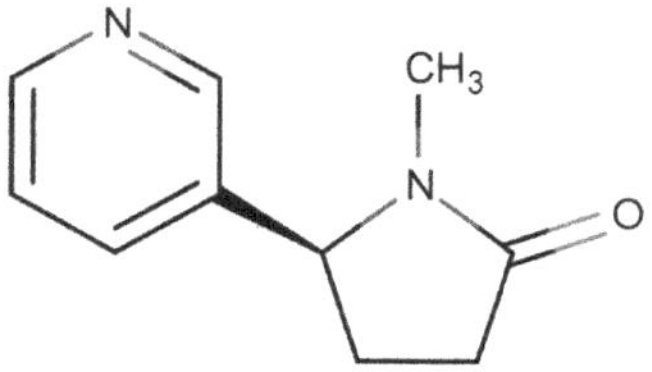

FIG. 8 Chemical structure of cotinine.

possess positive features similar to nicotine and may be effective in the treatment of AD, schizophrenia, and depression.

Cotinine improves information processing, attention, and memory-related task performance in model systems with relevance to both AD and other neuropsychiatric disorders such as schizophrenia.[495] Moreover, cotinine is cytoprotective and neuroprotective *in vitro*.[495–497] Cotinine also improved working (short term) memory in a delayed match-to-sample task in monkeys[496] and prevented memory loss in an AD mouse model, where its effects were accompanied by increased expression of the active form of PKB and postsynaptic density protein 95 in the hippocampus and frontal cortex of Tg6799 mice.[498] Cotinine acts via nAChRs exhibiting weak nicotinic agonism to desensitize receptors.[499] O'Leary and colleagues performed competition binding studies in a [3H]DA release test on medial and lateral caudate of the monkey brain and demonstrated that cotinine stimulated both nAChRs containing the α4 and β2 subunits and nAChRs composed of the α3 or α6 subunits and β2. Further, a functional discrimination of receptors by cotinine was demonstrated.[500] Cotinine stimulated Akt signaling, including inhibition of glycogen synthase kinase-3β (GSK-3β), which promoted neuronal survival and synaptic plasticity processes underlying learning and memory in the hippocampus and cortex of wild-type and Tg6799 AD mice.[501] More recently, Grizzell et al. demonstrated that cotinine exhibited antidepressant-like properties in mice subjected to prolonged restraint (a chronic stress model), and reduced cognitive impairment and synaptic loss in the hippocampus and prefrontal cortex through a mechanism involving preservation of brain homeostasis and expression of critical growth factors such as VEGF.[502] Encouragingly, cotinine offers some advantages over nicotine, including a much better safety profile,[503] much longer half-life,[504] and lower risk of abuse.[505] Hence, it may serve as a prototypical therapeutic agent for psychiatric disorders.

10.2.8 Ferulic Acid

Ferulic acid (Fig. 9) is a phenolic compound present in a variety of plants, including tobacco, which exerts antiinflammatory properties when topically administered in tetradecanoylphorbol acetate-induced mouse ear edema.[506]

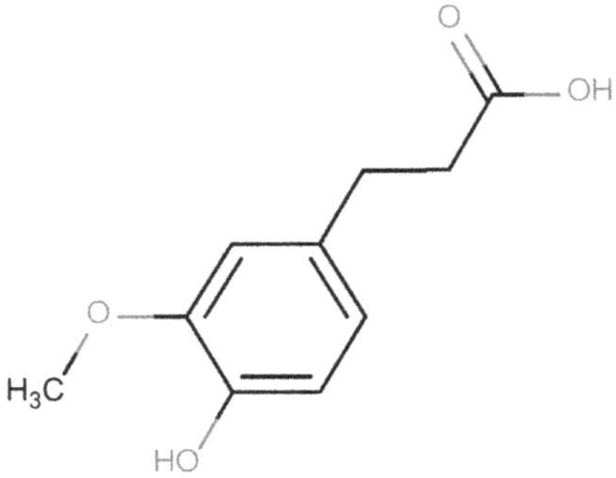

FIG. 9 Chemical structure of ferulic acid.

Yan and colleagues also showed that long-term administration of ferulic acid protects against α-amyloid peptide toxicity in the brain in a neurodegenerative mouse model and concluded that ferulic acid may be a useful chemopreventive agent against AD if treatment is initiated before neuroinflammatory processes develop.[507] Similar conclusions were reached in another study, in which ferulic acid was administered in an animal model of chronic neuroinflammation and showed a dose-dependent reduction in microglia activation within the temporal lobe.[508]

10.2.9 Quinoline

Quinoline (Fig. 10) is the most abundant aza-arene in cigarette smoke.[509] Quinoline is often part of the backbone of scaffolds that have been utilized for multiple purposes, not only principally in fertilizers but also notably as a past preventative measure for malarial infections.[510]

In the context of therapies for AD, Fiorito et al. investigated a series of quinoline derivatives that act as phosphodiesterase type 5 inhibitors. One of the compounds reversed synaptic and memory deficits in the APP/PS1 mouse model.[511] Further, an 8-hydroxyquinoline analog markedly decreased soluble interstitial brain Aβ within hours and improved cognitive performance in two types of amyloid-bearing transgenic mouse models of AD.[512]

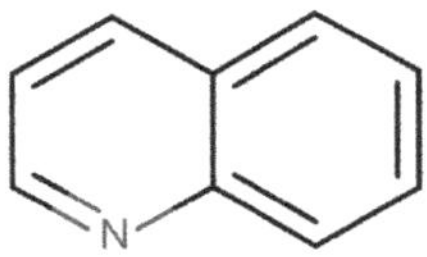

FIG. 10 Chemical structure of quinoline.

10.2.10 Resorcinol

Resorcinol (or resorcin) is a benzenediol (*m*-dihydroxybenzene) (Fig. 11). The resorcinol moiety has been found in a wide variety of natural products. In particular, plant phenolics, of which resorcinol ring-containing constituents are a part, are ubiquitous in nature. Resorcinol has also been found in tobacco leaves.[513] Topically, resorcinol is used to treat acne, seborrheic dermatitis, eczema, psoriasis, and other skin disorders. It is also used to treat corns, calluses, and warts. In a DSS-induced ulcerative colitis model, resorcinol administration significantly attenuated disease symptoms (reduced rectal bleeding and diarrhea, reduction of weight loss, and colon shortening). A reduction in pathology in DSS-treated resorcinol-injected mice, compared with DSS-treated vehicle-injected mice, was reported.[514]

Evidence for a role for resorcinol in neuroprotection has not been found. However, resorcinol moiety-containing compounds, such as cannabigerol, alleviate neuroinflammation in a chronic model of MS.[515]

10.2.11 Rutin

Rutin, also termed rutoside, quercetin-3-*O*-rutinoside, and sophorin, is the glycoside composed of flavonol quercetin and disaccharide rutinose (Fig. 12). Rutin is one of the major polyphenolic components of tobacco. Rutin inhibits Aβ aggregation and cytotoxicity, attenuates oxidative stress, and decreases *in vitro* production of NO and proinflammatory cytokines.[516]

Neuroprotective properties of rutin (or rutin metabolites) were demonstrated in a number of *in vivo* and *in vitro* studies. Aged rats pretreated with rutin (100 or 200 mg/kg) show altered dopamine and noradrenaline levels in the brain, coupled with improved spatial memory.[517] Small doses of rutin (10 mg/kg) demonstrate memory-enhancing effects using a step-through passive avoidance task in pentylenetetrazole-kindled rats.[518] Koda et al. examined prolonged rutin administration on trimethyltin-induced spatial memory impairment in rats and found that rutin significantly reduced memory impairments and pyramidal neuron damage in the hippocampal CA3b region. The authors suggested that rutin exerted an antioxidative effect.[519] Further, a protective effect of orally administered rutin (25 mg/kg of body weight, 3 weeks prior to 6-hydroxydopamine

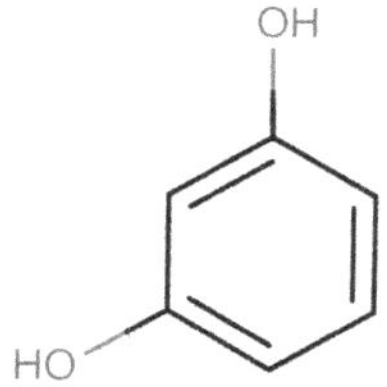

OH

HO

FIG. 11 Chemical structure of resorcinol.

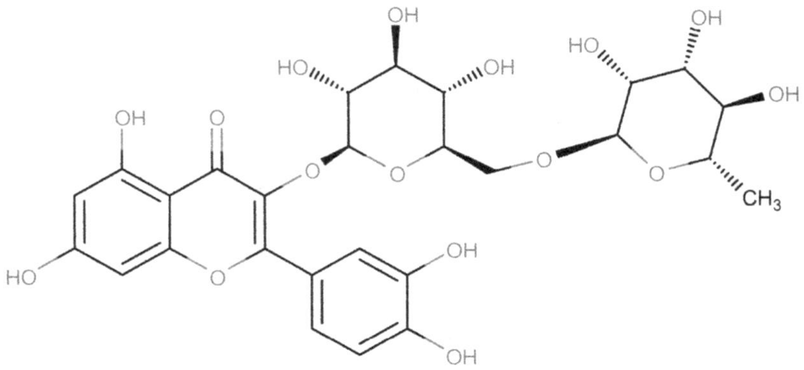

FIG. 12 Chemical structure of rutin.

injection) on dopaminergic neurons in a 6-hydroxydopamine-induced PD rat model was reported by Khan et al. In that study, rutin attenuated ischemic neural apoptosis by reducing p53 expression, preventing morphological changes, and increasing endogenous antioxidant enzymatic activities.[520] Rutin is an efficient natural ROS scavenger. Accordingly, rutin has been shown to protect pheochromocytoma (PC12) cells against 6-hydroxydopamine-induced neurotoxicity by improving antioxidant activity of SOD, catalase, and glutathione peroxidase and increasing total glutathione while reducing lipid peroxidation.[521] In a similar manner, rutin was shown to suppress ROS generation in H_2O_2-treated cells overexpressing the APP Swedish mutation.[522] After reporting that rutin inhibits Aβ aggregation and cytotoxicity, attenuates oxidative stress, and decreases production of NO and proinflammatory cytokines in SH-SY5Y neuroblastoma cells,[523] the same group found beneficial effects of orally administered rutin on spatial memory, Aβ burden, oxidative stress, and neuroinflammation in an APPswe/PS1dE9 transgenic mouse model.[516]

10.2.12 Scopoletin

Scopoletin (or 6-methoxy-7-hydroxycoumarin) (Fig. 13), identified in the roots of plants from the *Scopolia* genus, belongs to a class of chemicals related to coumarin. It was found in tobacco by Yang et al. in the late 1950s.[524]

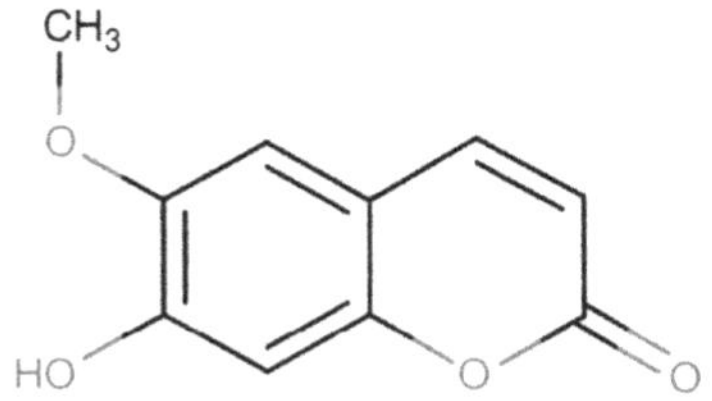

FIG. 13 Chemical structure of scopoletin.

Multiple effects of scopoletin isolated from various plants have been reported: antiinflammatory,[525,526] antiproliferative,[527] inhibition of inducible NO synthase[528,529] and prostaglandin synthase.[530] Scopoletin also inhibits MAO at moderate concentrations[531] and may act as an antioxidant[532] and radical scavenger.[533]

Inhibitory acetylcholinesterase (AChE) activity has also been reported,[534] which is particularly interesting in the context of AD. Hornick et al. demonstrated that scopoletin displays cholinergic transmission-enhancing and cognition and memory-improving properties, hypothesizing that its mode of action is via nAChRs.[535]

10.2.13 Summary

Several nonnicotine tobacco compounds with chemical structures similar to that of nicotine act or are proposed to act via various nAChRs to exert their antiinflammatory and neuroprotective effects and appear to exhibit properties similar to those of nicotine at the molecular level. Nevertheless, given the complexity of the nAChR subunit combinations that offer multiple modalities of function (either via direct receptor activation or modulation of their function) and differing distribution in the CNS and at the periphery, it is likely that additional, multifunctional properties of those compounds are involved, via either nAChRs or independently.

Other compounds have produced no evidence of action via nAChRs, and their effects are possibly exerted via inhibition of MAOs, reduction of inflammatory mediators, or other molecular mechanisms. Further research is required to characterize those mechanisms of action.

The list of nonnicotine tobacco compounds possibly responsible for the neuroprotective effects presented here is by no means exhaustive. Other less-investigated compounds may exert powerful effects that are yet to be described.

Chapter 11

Research Models of Neurodegenerative Diseases: Major Considerations for Translatability

The scientific evidence regarding potential neuroprotective effects of the tobacco-derived compounds mentioned above has been obtained from different experimental model systems that ideally should mimic human pathological processes. Although the process of understanding what causes different neuronal pathologies is still ongoing, scientific evidence on the underlying molecular mechanisms that lead to psychiatric and neurodegenerative diseases has accumulated over the past decades. Identification of mutations in familial forms of diseases has helped the development of a large number of genetically modified cell and animal models. When thoroughly characterized, these models are critical tools in the discovery and development of drugs that may efficiently prevent, slow down, and ideally reverse pathological processes. So far, significant advances and new insights have been provided, for example, on the mechanisms governing the pathological aggregation of key proteins, the nature and processes of neuronal damage, the role of genetic determinants, and the contribution of neuroinflammation in neuronal loss.

However, the use of these models appears to have shed light only on partial aspects of the various disorders, preventing true translation into new treatments, diagnostics, and prevention. The impressive amount of knowledge generated by experimental models has only marginally enriched the number of safe and efficient drugs. In fact, although preclinical results are often encouraging, when the new strategy is tested in the clinical setting, especially in large controlled clinical trials, the results are often disappointing. For example, very few treatments are currently approved for AD, and even those do not work for every patient. Of 244 compounds tested in 413 clinical trials between 2012 and 2014, only a combination of the already approved drugs donepezil and memantine was approved.[536,537]

One of the major shortcomings of the available models is that the pathologies they capture usually represent only part of the complex, incompletely

Nicotine and Other Tobacco Compounds in Neurodegenerative and Psychiatric Diseases.

https://doi.org/10.1016/B978-0-12-812922-7.00011-1

understood human disease. Even though complex transgenic animal models that capture the most important aspects of neuropathologies are developed, they fail to encompass all of the alterations of human neurodegeneration. In AD and PD research, for example, neuronal loss is apparent in only some of the models used. One possible improvement strategy would be the humanization of entire pathways in animal models and functional annotation of every gene in the context of whole organisms and different environments.[538] Introduction of human mutated genes has already been done for a number of AD models that are commercially available. Continuous efforts in that direction will enable the accumulation of knowledge that will aid in distinguishing cause from effect in lesions and lead to more sophisticated models bearing multiple transgenes mimicking more closely the complexity of the human disorder. To understand the relevance of the interactions among different neural cells and circuits, better models must mimic the human diseases more closely.

Adverse effects observed in clinical trials are another reason for the unfavorable outcome of therapeutic candidates that demonstrated promising results in animal models. One example is immunotherapy against Aβ, which showed some success in APP transgenic mice, yet the clinical trial had to be discontinued owing to the occurrence of meningoencephalitis in a subset of patients.[539] Encephalitis was not predicted by the mouse models, and only later was this side effect reported to occur in mice under rare circumstances. This lack of translation from rodents to humans may be due to insufficient depth of preclinical data analysis and suggests that better protocols for animal studies (taking into consideration the number of animals, sex, and general health conditions) should be implemented. Additionally, results obtained from preclinical experiments are often overinterpreted and the limitations of the models understated, while predictive validity is overstated, a possible reason for failure in clinical trials.

Similarly to other animal models of human pathologies, none of the EAE models represent human MS. EAE has become an instrumental player in MS research and has directly led to the development of three medications approved for MS: glatiramer acetate, mitoxantrone, and natalizumab. However, the EAE model has been criticized because numerous candidates that showed promising results in this mouse model ultimately either turned out to the lack of efficacy or in some cases cause harm in human MS.[540] The main reason for this gap is the use of immunologically immature, microbiologically clean, and genetically homogeneous rodent strains. Additionally, EAE has been proposed to be a $CD4^+$ T-cell-mediated autoimmune process, while analysis of human MS lesions revealed that the predominant immune cell types are $CD8^+$ T cells and macrophages, with $CD4^+$ T cells being much less frequent. Humanization of various rodent strains is a highly promising approach. Some of those models already exist, such as mouse models expressing single or multiple human transgenes.[541]

Although various *in vivo* models are currently in use, *in vitro* models may provide important insights into the pathogenesis of these disorders and represent an interesting approach for screening potential pharmacological agents. Today, by using human-induced pluripotent stem cells (hiPSCs), it has become possible to produce neurons and structures of the CNS[542] carrying the precise constellation of genetic variants that caused neurodegeneration in a given individual, which opens new opportunities for examining molecular processes that animal models or postmortem studies on human material may not offer. This "neurodegeneration-in-a-dish" approach holds substantial promise for gathering new insights into diseases of the nervous system. Furthermore, the use of multiple neuronal subtypes derived from large cohorts of patient-specific pluripotent cell lines make identification of resistant and vulnerable neurons in different pathologies possible. Careful studies of the immediate changes induced in the most sensitive neuronal types may provide insight with therapeutic value, long before end points of degeneration such as protein inclusion formation or neuronal death are reached. In addition, perturbation of pathophysiological processes *in vitro* can be studied in more manageable time frames. However, the complex disease etiology, the different time scales of neurodegeneration in humans versus *in vitro* studies, and the variability of the results obtained depending on the deployed reprogramming methodology pose particular challenges to stem-cell-based models. These cells may not truly reflect the cellular responses to compounds that the body would have at a physiological level. All that makes translation to the clinic limited.[543] Although hiPSCs technology is in its infancy and faces many obstacles, it offers great potential in helping to identify therapeutic targets for treating neurodegenerative diseases. As more defined disease-specific molecular signatures are uncovered and *in vitro* maturation of stem-cell-derived neurons becomes more predictable, better models capable of characterizing previously unknown disease mechanisms will be established.

Chapter 12

Concluding Remarks

Smoking is harmful to human health and causes a number of serious diseases. To avoid the health risks associated with smoking, one should never start, and for smokers, the best way to reduce these risks is to quit. Mechanistic understanding of the impact of tobacco smoke and the constituents of smoke on critical biological processes has been a topic of research for the scientific community for a long time and is far from being completely elucidated. But, the potential neuroprotective effects of nicotine and other compounds may be of interest if their use can be isolated from that of cigarette smoking and its associated harms. Further identification of those compounds and investigations into their modes of action may offer possible explanations for the intriguing results from epidemiological studies for some neurodegenerative diseases such as PD.

In this context, nicotine is clearly the most studied tobacco compound, and much knowledge has been accumulated on its pharmacological effects. Nicotine exerts a multitude of effects either via activation of nAChRs or through other mechanisms. There are many apparent contradictions in the studies on nicotine in neurodegenerative diseases, which may be explained by differences in dose, route of administration, or other aspects, but as of yet, these are still not fully understood. Pharmaceutical companies have invested significant efforts in developing different agonists and modulators of nAChRs, with often disappointing outcomes. Prolonged activation of nAChRs or their desensitization by long-term nicotine administration can be neurotoxic to developing neurons or, at high doses, to adult neurons. Nicotine can increase oxidative stress at high concentrations (1 and 10 mM), while at lower concentrations (10 μM), it reduced lipid peroxidation and increased viability of PC12 rat pheochromocytoma cells.[544] Possible mechanisms of nicotine-mediated neuroprotection involve the cholinergic system (desensitization), stimulation of dopamine release, upregulation of prosurvival molecular pathways (neuroprotection), attenuation of inflammation and modulation of the immune response, and the interplay between membrane and mitochondrial nAChR actions. It is possible that research into the role of nicotine in such diseases has been limited by concerns around the addictive nature of nicotine. Nevertheless, nicotine is the subject of several ongoing clinical trials that indicate that the potentials of this powerful alkaloid still attract quite some attention.

Results of clinical trials with nicotine are mixed and do not entirely reflect the observations from epidemiological studies on smoking. The route of

Nicotine and Other Tobacco Compounds in Neurodegenerative and Psychiatric Diseases.

https://doi.org/10.1016/B978-0-12-812922-7.00012-3

administration may be one of the reasons for such an outcome, as differences between pharmacologically active doses of nicotine that is inhaled and nicotine administered as gum, patches, or lozenges are to be expected. Disease stage may also contribute to variability and inconsistency.

Finally, nicotine may not be the only component of smoke responsible for these effects. Other alkaloids such as anatabine and anabasine, MAO inhibitors, cembranoids, and others may also be involved in neuroprotective effects through a complex mode of action. Those compounds should be strictly decoupled from smoke constituents and further studied to understand their mechanisms of actions in detail. These individual compounds are interesting and may have pharmaceutical potential, similarly to other medicinal plant alkaloids that have been used in various cultures for centuries. Such potential makes these compounds worth focusing on, either individually or in various combinations.

Abbreviations

5-HT	5-hydroxytryptamine
6-OHDA	6-hydroxydopamine
Aβ	amyloid beta
Ab	antibody
ACC	anterior cingulate cortex
ACh	acetylcholine
AChE	acetylcholinesterase
AD	Alzheimer's disease
ADHD	attention deficit hyperactivity disorder
Akt	serine/threonine protein kinase
APOE	apolipoprotein E
APP	amyloid precursor protein
ATD	acute tryptophan depletion
β- CTF	beta C-terminal fragment
BBB	blood-brain barrier
Bcl-2	B-cell lymphoma 2
Bcl-x	B-cell lymphoma 2-like
bFGF-2	basic fibroblast growth factor-2
b.i.d.	bis in die (twice per day)
CaMKII	calcium effector protein calmodulin
cAMP	cyclic adenosine monophosphate
cig	cigarettes
CI	confidence interval
CNS	central nervous system
COMT	catechol-*o*-methyltransferase
COPD	chronic obstructive pulmonary disease
COX-2	cyclooxygenase 2
CREB	cAMP-response element binding
CS	cigarette smoke
DA	dopamine
DAOA	D-amino acid oxidase activator
DAT	dopamine transporter
DISC1	disrupted-in-schizophrenia 1
DOI	(1-)2,5-dimethoxy-4-iodophenyl-2-aminopropane
DRD4	dopamine receptor D4
DSS	dextran sulfate sodium
DTNBP1	dystrobrevin-binding protein 1
EAE	experimental autoimmune encephalomyelitis
EGF	epidermal growth factor
Egr-1	early growth response 1
Egr-2	early growth response 2

ERK/MAPK	extracellular signal-regulated mitogen-activated protein kinase
FDA	Food and Drug Administration
FRL	Flinders resistant line
FSL	Flinders sensitive line
GABA	γ-aminobutyric acid
GC-MS	gas chromatography mass spectrometry
GSK 3β	glycogen synthase kinase 3β
hiPSC	human-induced pluripotent stem cells
HLA-DRB1	major histocompatibility complex, class II, DR beta 1
HPLC	high-performance liquid chromatography
IC_{50}	half-maximal inhibitory concentration
ICV	intracerebroventricular
IGF-I	insulin-like growth factor I
IL-6	interleukin 6
i.p.	intraperitoneal
IUPHAR	International Union of Basic and Clinical Pharmacology
JAK2	Janus kinase 2
JNK	Jun N-terminal kinase
KO	knockout
L-DOPA	levodopa
LGP	lateral globus pallidus
LHb	lateral habenula
LPS	lipopolysaccharide
LTP	long-term potentiation
MAO	monoamine oxidase
MCI	mild cognitive impairment
MOG	myelin oligodendrocyte glycoprotein
MPTP	1-methyl-4-phenyl-1,2,3,6-tetrahydropyridine
MS	multiple sclerosis
NA	not applicable
NAc	nucleus accumbens
nAChR	nicotinic acetylcholine receptor
NADH	reduced form of nicotinamide adenine dinucleotide
NAT1	*N*-acetyltransferase 1
NCT	National Clinical Trial
NFκB	nuclear factor kappa B
NFTs	neurofibrillary tangles
NGF	nerve growth factor
NMDA	*N*-methyl-D-aspartate
NNK	4-(methylnitrosamino)-1-(3-pyridyl)-1-butanone
NO	nitric oxide
NPC	neural progenitor cell
NPY	neuropeptide Y
NRG1	neuregulin 1
NRT	nicotine replacement therapy
Nurr77	nuclear orphan receptor 77
OCD	obsessive-compulsive disorder
OR	odds ratio

PCP	phencyclidine
PD	Parkinson's disease
PET	positron emission tomography
PI3K	phosphatidylinositol 3-kinase
PKA	protein kinase A
PKB	protein kinase B
PKC	protein kinase C
PnC	pontine reticular nucleus
PPI	prepulse inhibition
PPMS	primary-progressive multiple sclerosis
PRODH	proline dehydrogenase
PSEN1	presenilin 1
PSEN2	presenilin 2
RAS/RAF/MEK/ERK	signaling networks that govern proliferation, differentiation, and cell survival
REM	random eye movement
ROS	reactive oxygen species
RR	relative risk
RRMS	relapsing-remitting multiple sclerosis
RT	response time
sAPPα	soluble extracellular N-terminal fragment of amyloid precursor protein
s.c.	subcutaneous
SLITRK1	SLIT and NTRK-like family member 1
SNAP	*S*-nitroso-*N*-acetylpenicillamine
SOCS3	suppressor of cytokine signaling 3
SOD	superoxide dismutase
SPMS	secondary-progressive multiple sclerosis
Src	Src proto-oncogene, nonreceptor tyrosine kinase
STAT3	signal transducer and activator of transcription 3
TAAR6	trace amine-associated receptor 6
TH	tyrosine hydroxylase
TNF	tumor necrosis factor
trkB	tropomyosin receptor kinase B
TS	Tourette's syndrome
VEGF	vascular endothelial growth factor
ZDHHC8	zinc-finger DHHC type-containing 8

References

1. Daly JW, Garraffo HM, Spande TF, Decker MW, Sullivan JP, Williams M. Alkaloids from frog skin: the discovery of epibatidine and the potential for developing novel non-opioid analgesics. *Nat Prod Rep* 2000;**17**:131–5.
2. Marks MJ, Laverty DS, Whiteaker P, et al. John Daly's compound, epibatidine, facilitates identification of nicotinic receptor subtypes. *J Mol Neurosci* 2010;**40**:96–104.
3. Yin J, Chen W, Yang H, Xue M, Schaaf CP. Chrna7 deficient mice manifest no consistent neuropsychiatric and behavioral phenotypes. *Sci Rep* 2017;**7**:39941.
4. Lewis AS, van Schalkwyk GI, Bloch MH. Alpha-7 nicotinic agonists for cognitive deficits in neuropsychiatric disorders: a translational meta-analysis of rodent and human studies. *Prog Neuropsychopharmacol Biol Psychiatry* 2017;**75**:45–53.
5. Borovikova LV, Ivanova S, Zhang M, et al. Vagus nerve stimulation attenuates the systemic inflammatory response to endotoxin. *Nature* 2000;**405**:458–62.
6. Pavlov VA, Wang H, Czura CJ, Friedman SG, Tracey KJ. The cholinergic anti-inflammatory pathway: a missing link in neuroimmunomodulation. *Mol Med* 2003;**9**:125–34.
7. Carter BD, Freedman ND, Jacobs EJ. Smoking and mortality – beyond established causes. *N Engl J Med* 2015;**372**:2170.
8. U.S. Department of Health and Human Services. *The health consequences of smoking—50 years of progress. A report of the surgeon general.* Atlanta, GA; 2014.
9. Russell MA. Low-tar medium-nicotine cigarettes: a new approach to safer smoking. *Br Med J* 1976;**1**:1430–3.
10. McNeil A. Reducing harm from nicotine use. *Fifty years since smoking and health progress, lessons and priorities for a smoke-free UK.* London: Royal College of Physicians; 2012.
11. Royal College of Physicians Report. *Nicotine without smoke: tobacco harm reduction. A report by the tobacco advisory group of the royal college of physicians.* London: RCP; 2016.
12. Waldum HL, Nilsen OG, Nilsen T, et al. Long-term effects of inhaled nicotine. *Life Sci* 1996;**58**:1339–46.
13. Quik M. Smoking, nicotine and Parkinson's disease. *Trends Neurosci* 2004;**27**:561–8.
14. Weston M, Constantinescu CS. What role does tobacco smoking play in multiple sclerosis disability and mortality? A review of the evidence. *Neurodegener Dis Manag* 2015;**5**:19–25.
15. Hedström AK, Bäärnhielm M, Olsson T, Alfredsson L. Tobacco smoking, but not Swedish snuff use, increases the risk of multiple sclerosis. *Neurology* 2009;**73**:696–701.
16. Peters R, Poulter R, Warner J, Beckett N, Burch L, Bulpitt C. Smoking, dementia and cognitive decline in the elderly, a systematic review. *BMC Geriatr* 2008;**8**:36.
17. Cataldo JK, Prochaska JJ, Glantz SA. Cigarette smoking is a risk factor for Alzheimer's disease: an analysis controlling for tobacco industry affiliation. *J Alzheimers Dis* 2010;**19**:465–80.
18. Newhouse PA, Sunderland T, Tariot PN, et al. Intravenous nicotine in Alzheimer's disease: a pilot study. *Psychopharmacology* 1988;**95**:171–5.

19. Jones G, Sahakian B, Levy R, Warburton DM, Gray J. Effects of acute subcutaneous nicotine on attention, information processing and short-term memory in Alzheimer's disease. *Psychopharmacology* 1992;**108**:485–94.
20. McClernon FJ, Kollins SH. ADHD and smoking. *Ann N Y Acad Sci* 2008;**1141**:131–47.
21. Boden JM, Fergusson DM, Horwood LJ. Cigarette smoking and depression: tests of causal linkages using a longitudinal birth cohort. *Br J Psychiatry* 2010;**196**:440–6.
22. Quik M, Perez XA, Bordia T. Nicotine as a potential neuroprotective agent for Parkinson's disease. *Mov Disord* 2012;**27**:947–57.
23. Jonge W, Ulloa L. The alpha7 nicotinic acetylcholine receptor as a pharmacological target for inflammation. *Br J Pharmacol* 2007;**151**:915–29.
24. Copeland Jr. RL, Das JR, Kanaan YM, Taylor RE, Tizabi Y. Antiapoptotic effects of nicotine in its protection against salsolinol-induced cytotoxicity. *Neurotox Res* 2007;**12**:61–9.
25. Picciotto MR, Addy NA, Mineur YS, Brunzell DH. It is not "either/or": activation and desensitization of nicotinic acetylcholine receptors both contribute to behaviors related to nicotine addiction and mood. *Prog Neurobiol* 2008;**84**:329–42.
26. Benowitz NL. Nicotine addiction. *N Engl J Med* 2010;**362**:2295.
27. Balfour DJ. The neuronal pathways mediating the behavioral and addictive properties of nicotine. *Handb Exp Pharmacol* 2009;**192**:209–33.
28. Caggiula AR, Donny EC, Chaudhri N, Perkins KA, Evans-Martin FF, Sved AF. Importance of nonpharmacological factors in nicotine self-administration. *Physiol Behav* 2002;**77**:683–7.
29. U.S. Food and Drug Administration. *Nicotine replacement therapy labels may change*; 2013.
30. Fagerstrom K, Eissenberg T. Dependence on tobacco and nicotine products: a case for product-specific assessment. *Nicotine Tob Res* 2012;**14**:1382–90.
31. Mayer B. How much nicotine kills a human? Tracing back the generally accepted lethal dose to dubious self-experiments in the nineteenth century. *Arch Toxicol* 2014;**88**:5–7.
32. Niaura R. Nicotine in addictions: a comprehensive guidebook. *Addictions: a comprehensive guidebook.* 2nd ed. Oxford University Press, Oxford, UK; 2013.
33. Stepanov I, Jensen J, Hatsukami D, Hecht SS. New and traditional smokeless tobacco: comparison of toxicant and carcinogen levels. *Nicotine Tob Res* 2008;**10**:1773–82.
34. Baba S, Wikstrom AK, Stephansson O, Cnattingius S. Changes in snuff and smoking habits in Swedish pregnant women and risk for small for gestational age births. *BJOG* 2013;**120**:456–62.
35. England LJ, Levine RJ, Mills JL, Klebanoff MA, Yu KF, Cnattingius S. Adverse pregnancy outcomes in snuff users. *Am J Obstet Gynecol* 2003;**189**:939–43.
36. Wikstrom AK, Cnattingius S, Stephansson O. Maternal use of Swedish snuff (snus) and risk of stillbirth. *Epidemiology* 2010;**21**:772–8.
37. Button TM, Maughan B, McGuffin P. The relationship of maternal smoking to psychological problems in the offspring. *Early Hum Dev* 2007;**83**:727–32.
38. Tiesler CM, Heinrich J. Prenatal nicotine exposure and child behavioural problems. *Eur Child Adolesc Psychiatry* 2014;**23**:913–29.
39. Schapira AH, Jenner P. Etiology and pathogenesis of Parkinson's disease. *Mov Disord* 2011;**26**:1049–55.
40. Schapira AH. Neurobiology and treatment of Parkinson's disease. *Trends Pharmacol Sci* 2009;**30**:41–7.
41. Parkinson Disease Foundation. *Statistics on Parkinson's*; 2016.
42. Hobson P, Gallacher J, Meara J. Cross-sectional survey of Parkinson's disease and parkinsonism in a rural area of the United Kingdom. *Mov Disord* 2005;**20**:995–8.

43. Quik M, Wonnacott S. α6β2* and α4β2* nicotinic acetylcholine receptors as drug targets for Parkinson's disease. *Pharmacol Rev* 2011;**63**:938–66.
44. Cheung ZH, Ip NY. The emerging role of autophagy in Parkinson's disease. *Mol Brain* 2009;**2**:1.
45. McGeer PL, McGeer EG. Inflammation and neurodegeneration in Parkinson's disease. *Parkinsonism Relat Disord* 2004;**10**:S3–7.
46. Lee MK, Stirling W, Xu Y, et al. Human α-synuclein-harboring familial Parkinson's disease-linked Ala-53 → Thr mutation causes neurodegenerative disease with α-synuclein aggregation in transgenic mice. *Proc Natl Acad Sci U S A* 2002;**99**:8968–73.
47. Saracchi E, Fermi S, Brighina L. Emerging candidate biomarkers for Parkinson's disease: a review. *Aging Dis* 2014;**5**:27–34.
48. Olanow CW, Brundin P. Parkinson's disease and alpha synuclein: is Parkinson's disease a prion-like disorder? *Mov Disord* 2013;**28**:31–40.
49. Milber JM, Noorigian JV, Morley JF, et al. Lewy pathology is not the first sign of degeneration in vulnerable neurons in Parkinson disease. *Neurology* 2012;**79**:2307–14.
50. Eugster L, Bargiotas P, Bassetti CL, Michael Schuepbach WM. Deep brain stimulation and sleep-wake functions in Parkinson's disease: a systematic review. *Parkinsonism Relat Disord* 2016;**32**:12–9.
51. Allam MF, Campbell MJ, Hofman A, Del Castillo AS, Fernández-Crehuet Navajas R. Smoking and Parkinson's disease: systematic review of prospective studies. *Mov Disord* 2004;**19**:614–21.
52. Gorell JM, Rybicki BA, Johnson CC, Peterson EL. Smoking and Parkinson's disease: a dose–response relationship. *Neurology* 1999;**52**:115.
53. Baron JA. Beneficial effects of nicotine and cigarette smoking: the real, the possible and the spurious. *Br Med Bull* 1996;**52**:58–73.
54. Morens D, Grandinetti A, Reed D, White L, Ross G. Cigarette smoking and protection from Parkinson's disease false association or etiologic clue? *Neurology* 1995;**45**:1041–51.
55. O'Reilly EJ, McCullough ML, Chao A, et al. Smokeless tobacco use and the risk of Parkinson's disease mortality. *Mov Disord* 2005;**20**:1383–4.
56. Ritz B, Ascherio A, Checkoway H, et al. Pooled analysis of tobacco use and risk of Parkinson disease. *Arch Neurol* 2007;**64**:990–7.
57. Ross GW, Petrovitch H. Current evidence for neuroprotective effects of nicotine and caffeine against Parkinson's disease. *Drugs Aging* 2001;**18**:797–806.
58. Thacker E, O'Reilly E, Weisskopf M, et al. Temporal relationship between cigarette smoking and risk of Parkinson disease. *Neurology* 2007;**68**:764–8.
59. Alves G, Kurz M, Lie SA, Larsen JP. Cigarette smoking in Parkinson's disease: influence on disease progression. *Mov Disord* 2004;**19**:1087–92.
60. Driver J, Kurth T, Buring J, Gaziano J, Logroscino G. Parkinson disease and risk of mortality: a prospective comorbidity-matched cohort study. *Neurology* 2008;**70**:1423–30.
61. Van Der Mark M, Nijssen PC, Vlaanderen J, et al. A case-control study of the protective effect of alcohol, coffee, and cigarette consumption on Parkinson disease risk: time-since-cessation modifies the effect of tobacco smoking. *PLoS One* 2014;**9**:e95297.
62. Saaksjarvi K, Knekt P, Mannisto S, et al. Reduced risk of Parkinson's disease associated with lower body mass index and heavy leisure-time physical activity. *Eur J Epidemiol* 2014;**29**:285–92.
63. Ton TG, Biggs ML, Comer D, et al. Enhancing case ascertainment of Parkinson's disease using Medicare claims data in a population-based cohort: the Cardiovascular Health Study. *Pharmacoepidemiol Drug Saf* 2014;**23**:119–27.

64. O'Reilly EJ, McCullough ML, Chao A, et al. Smokeless tobacco use and the risk of Parkinson's disease mortality. *Mov Disord* 2005;**20**:1383–4.
65. Benedetti MD, Bower JH, Maraganore DM, et al. Smoking, alcohol, and coffee consumption preceding Parkinson's disease: a case-control study. *Neurology* 2000;**55**:1350–8.
66. Yang F, Pedersen NL, Ye W, et al. Moist smokeless tobacco (snus) use and risk of Parkinson's disease. *Int J Epidemiol* 2016;**46**:872–80.
67. Rapier C, Lunt GG, Wonnacott S. Nicotinic modulation of [3H] dopamine release from striatal synaptosomes: pharmacological characterisation. *J Neurochem* 1990;**54**:937–45.
68. Lichtensteiger W, Hefti F, Felix D, Huwyler T, Melamed E, Schlumpf M. Stimulation of nigrostriatal dopamine neurones by nicotine. *Neuropharmacology* 1982;**21**:963–8.
69. Westfall TC, Besson M-J, Giorguieff M-F, Glowinski J. The role of presynaptic receptors in the release and synthesis of 3H-dopamine by slices of rat striatum. *Naunyn Schmiedeberg's Arch Pharmacol* 1976;**292**:279–87.
70. Zhou FM, Wilson CJ, Dani JA. Cholinergic interneuron characteristics and nicotinic properties in the striatum. *J Neurobiol* 2002;**53**:590–605.
71. O'Neill M, Murray T, Lakics V, Visanji N, Duty S. The role of neuronal nicotinic acetylcholine receptors in acute and chronic neurodegeneration. *Curr Drug Targets CNS Neurol Disord* 2002;**1**:399–411.
72. Ward RJ, Lallemand F, De Witte P, Dexter DT. Neurochemical pathways involved in the protective effects of nicotine and ethanol in preventing the development of Parkinson's disease: potential targets for the development of new therapeutic agents. *Prog Neurobiol* 2008;**85**:135–47.
73. Quik M, Kulak JM. Nicotine and nicotinic receptors; relevance to Parkinson's disease. *Neurotoxicology* 2002;**23**:581–94.
74. Quik M, O'Neill M, Perez XA. Nicotine neuroprotection against nigrostriatal damage: importance of the animal model. *Trends Pharmacol Sci* 2007;**28**:229–35.
75. Court JA, Martin-Ruiz C, Graham A, Perry E. Nicotinic receptors in human brain: topography and pathology. *J Chem Neuroanat* 2000;**20**:281–98.
76. Guan ZZ, Nordberg A, Mousavi M, Rinne JO, Hellstrom-Lindahl E. Selective changes in the levels of nicotinic acetylcholine receptor protein and of corresponding mRNA species in the brains of patients with Parkinson's disease. *Brain Res* 2002;**956**:358–66.
77. Bordia T, Grady SR, McIntosh JM, Quik M. Nigrostriatal damage preferentially decreases a subpopulation of alpha6beta2* nAChRs in mouse, monkey, and Parkinson's disease striatum. *Mol Pharmacol* 2007;**72**:52–61.
78. Liu Q, Emadi S, Shen JX, Sierks MR, Wu J. Human alpha4beta2 nicotinic acetylcholine receptor as a novel target of oligomeric alpha-synuclein. *PLoS One* 2013;**8**:e55886.
79. Alzheimer's Statistics. *2016 Alzheimer's statistics*; 2016.
80. Lambert JC, Ibrahim-Verbaas CA, Harold D, et al. Meta-analysis of 74,046 individuals identifies 11 new susceptibility loci for Alzheimer's disease. *Nat Genet* 2013;**45**:1452–8.
81. Reitz C, Mayeux R. Alzheimer disease: epidemiology, diagnostic criteria, risk factors and biomarkers. *Biochem Pharmacol* 2014;**88**:640–51.
82. Ferini-Strambi L, Smirne S, Garancini P, Pinto P, Franceschi M. Clinical and epidemiological aspects of Alzheimer's disease with presenile onset: a case control study. *Neuroepidemiology* 1990;**9**:39–49.
83. Tyas S. Are tobacco and alcohol use related to Alzheimer's disease? A critical assessment of the evidence and its implications. *Addict Biol* 1996;**1**:237–54.
84. Launer L, Feskens E, Kalmijn S, Kromhout D. Smoking, drinking, and thinking: the Zutphen Elderly Study. *Am J Epidemiol* 1996;**143**:219–27.
85. Merchant C, Tang M-X, Albert S, Manly J, Stern Y, Mayeux R. The influence of smoking on the risk of Alzheimer's disease. *Neurology* 1999;**52**:1408.

86. Ott A, Slooter A, Hofman A, et al. Smoking and risk of dementia and Alzheimer's disease in a population-based cohort study: the Rotterdam Study. *Lancet* 1998;**351**:1840–3.
87. Elias PK, Elias MF, Robbins MA, Budge MM. Blood pressure-related cognitive decline does age make a difference? *Hypertension* 2004;**44**:631–6.
88. Hebert LE, Scherr PA, Beckett LA, et al. Relation of smoking and alcohol consumption to incident Alzheimer's disease. *Am J Epidemiol* 1992;**135**:347–55.
89. Traber MG, van der Vliet A, Reznick AZ, Cross CE. Tobacco-related diseases: is there a role for antioxidant micronutrient supplementation? *Clin Chest Med* 2000;**21**:173–87.
90. Zhong G, Wang Y, Zhang Y, Guo JJ, Zhao Y. Smoking is associated with an increased risk of dementia: a meta-analysis of prospective cohort studies with investigation of potential effect modifiers. *PLoS One* 2015;**10**:e0118333.
91. Prince M, Albanese E, Guerchet M, Prina M. *World Alzheimer Report 2014. Dementia and risk reduction: an analysis of protective and modifiable factors*. Londres: Alzheimers Disease International; 2014.
92. Oz M, Lorke DE, Yang KH, Petroianu G. On the interaction of beta-amyloid peptides and alpha7-nicotinic acetylcholine receptors in Alzheimer's disease. *Curr Alzheimer Res* 2013;**10**:618–30.
93. Posadas I, Lopez-Hernandez B, Cena V. Nicotinic receptors in neurodegeneration. *Curr Neuropharmacol* 2013;**11**:298–314.
94. Lopez-Arrieta JM, Rodriguez JL, Sanz F. Efficacy and safety of nicotine on Alzheimer's disease patients. *Cochrane Database Syst Rev* 2001;**2**:CD001749.
94a. Howe MN, Price IR. Effects of transdermal nicotine on learning, memory, verbal fluency, concentration, and general health in a healthy sample at risk for dementia. Int Psychogeriatr 2001;**13**:465–75.
94b. Picciotto MR, Caldarone BJ, King SL, Zachariou V. Nicotinic receptors in the brain. Links between molecular biology and behavior. Neuropsychopharmacology 2000;**22**:451–65.
95. Newhouse P, Kellar K, Aisen P, et al. Nicotine treatment of mild cognitive impairment: a 6-month double-blind pilot clinical trial. *Neurology* 2012;**78**:91–101.
96. Buccafusco JJ, Jackson W, Jonnala R, Terry Jr. A. Differential improvement in memory-related task performance with nicotine by aged male and female rhesus monkeys. *Behav Pharmacol* 1999;**10**:681–90.
97. Rezvani AH, Levin ED. Cognitive effects of nicotine. *Biol Psychiatry* 2001;**49**:258–67.
98. Noshita T, Murayama N, Nakamura S. Effect of nicotine on neuronal dysfunction induced by intracerebroventricular infusion of amyloid-β peptide in rats. *Eur Rev Med Pharmacol Sci* 2015;**19**:334–43.
99. Kadir A, Darreh-Shori T, Almkvist O, et al. PET imaging of the in vivo brain acetylcholinesterase activity and nicotine binding in galantamine-treated patients with AD. *Neurobiol Aging* 2008;**29**:1204–17.
100. Nordberg A. PET studies and cholinergic therapy in Alzheimer's disease. *Rev Neurol (Paris)* 1999;**155**(Suppl. 4):S53–63.
101. Guan ZZ, Zhang X, Ravid R, Nordberg A. Decreased protein levels of nicotinic receptor subunits in the hippocampus and temporal cortex of patients with Alzheimer's disease. *J Neurochem* 2000;**74**:237–43.
102. Martin-Ruiz CM, Court JA, Molnar E, et al. Alpha4 but not alpha3 and alpha7 nicotinic acetylcholine receptor subunits are lost from the temporal cortex in Alzheimer's disease. *J Neurochem* 1999;**73**:1635–40.
103. Nordberg A, Winblad B. Reduced number of [3H]nicotine and [3H]acetylcholine binding sites in the frontal cortex of Alzheimer brains. *Neurosci Lett* 1986;**72**:115–9.

104. Hardy J, Selkoe DJ. The amyloid hypothesis of Alzheimer's disease: progress and problems on the road to therapeutics. *Science* 2002;**297**:353–6.
105. Ahmed M, Davis J, Aucoin D, et al. Structural conversion of neurotoxic amyloid-beta(1-42) oligomers to fibrils. *Nat Struct Mol Biol* 2010;**17**:561–7.
106. Pettit DL, Shao Z, Yakel JL. Beta-amyloid(1-42) peptide directly modulates nicotinic receptors in the rat hippocampal slice. *J Neurosci* 2001;**21**:RC120.
107. Liu Q, Kawai H, Berg DK. Beta -amyloid peptide blocks the response of alpha 7-containing nicotinic receptors on hippocampal neurons. *Proc Natl Acad Sci U S A* 2001;**98**:4734–9.
108. Dineley KT, Bell KA, Bui D, Sweatt JD. Beta -amyloid peptide activates alpha 7 nicotinic acetylcholine receptors expressed in Xenopus oocytes. *J Biol Chem* 2002;**277**:25056–61.
109. Ramsden M, Henderson Z, Pearson HA. Modulation of Ca2+ channel currents in primary cultures of rat cortical neurones by amyloid beta protein (1-40) is dependent on solubility status. *Brain Res* 2002;**956**:254–61.
110. Lilja AM, Porras O, Storelli E, Nordberg A, Marutle A. Functional interactions of fibrillar and oligomeric amyloid-beta with alpha7 nicotinic receptors in Alzheimer's disease. *J Alzheimers Dis* 2011;**23**:335–47.
111. Hellstrom-Lindahl E, Moore H, Nordberg A. Increased levels of tau protein in SH-SY5Y cells after treatment with cholinesterase inhibitors and nicotinic agonists. *J Neurochem* 2000;**74**:777–84.
112. Wang HY, Li W, Benedetti NJ, Lee DH. Alpha 7 nicotinic acetylcholine receptors mediate beta-amyloid peptide-induced tau protein phosphorylation. *J Biol Chem* 2003;**278**:31547–53.
113. Oddo S, Caccamo A, Green KN, et al. Chronic nicotine administration exacerbates tau pathology in a transgenic model of Alzheimer's disease. *Proc Natl Acad Sci U S A* 2005;**102**:3046–51.
114. Yin Y, Wang Y, Gao D, et al. Accumulation of human full-length tau induces degradation of nicotinic acetylcholine receptor alpha4 via activating calpain-2. *Sci Rep* 2016;**6**:27283.
115. McFarland HF, Martin R. Multiple sclerosis: a complicated picture of autoimmunity. *Nat Immunol* 2007;**8**:913–9.
116. Miljković D, Spasojević I. Multiple sclerosis: molecular mechanisms and therapeutic opportunities. *Antioxid Redox Signal* 2013;**19**:2286–334.
117. Goodin DS. The epidemiology of multiple sclerosis: insights to disease pathogenesis. *Handb Clin Neurol* 2014;**122**:231–66.
118. Hawker K. Progressive multiple sclerosis: characteristics and management. *Neurol Clin* 2011;**29**:423–34.
119. Sawcer S, Hellenthal G, Pirinen M, et al. Genetic risk and a primary role for cell-mediated immune mechanisms in multiple sclerosis. *Nature* 2011;**476**:214–9.
120. Van der Mei I, Simpson S, Stankovich J, Taylor B. Individual and joint action of environmental factors and risk of MS. *Neurol Clin* 2011;**29**:233–55.
121. Lassmann H, van Horssen J, Mahad D. Progressive multiple sclerosis: pathology and pathogenesis. *Nat Rev Neurol* 2012;**8**:647–56.
122. Baranzini SE, Hauser SL. Large-scale gene-expression studies and the challenge of multiple sclerosis. *Genome Biol* 2002;**3**:1027.
123. Prineas J, Connell F. Remyelination in multiple sclerosis. *Ann Neurol* 1979;**5**:22–31.
124. Cregg JM, DePaul MA, Filous AR, Lang BT, Tran A, Silver J. Functional regeneration beyond the glial scar. *Exp Neurol* 2014;**253**:197–207.
125. Mao P, Reddy PH. Is multiple sclerosis a mitochondrial disease? *Biochim Biophys Acta* 2010;**1802**:66–79 [molecular basis of disease].
126. Van Horssen J, Witte M, Ciccarelli O. The role of mitochondria in axonal degeneration and tissue repair in MS. *Mult Scler J* 2012;**18**:1058–67. https://doi.org/10.1177/1352458512452924.

127. Hedström AK, Hillert J, Olsson T, Alfredsson L. Smoking and multiple sclerosis susceptibility. *Eur J Epidemiol* 2013;**28**:867–74.
128. Briggs FB, Acuna B, Shen L, et al. Smoking and risk of multiple sclerosis: evidence of modification by NAT1 variants. *Epidemiology* 2014;**25**:605–14.
129. Hedström A, Olsson T, Alfredsson L. Smoking is a major preventable risk factor for multiple sclerosis. *Mult Scler J* 2015;**22**:1021–6. https://doi.org/10.1177/1352458515609794.
130. Shi F-D, Piao W-H, Kuo Y-P, Campagnolo DI, Vollmer TL, Lukas RJ. Nicotinic attenuation of central nervous system inflammation and autoimmunity. *J Immunol* 2009;**182**:1730–9.
131. Gao Z, Nissen JC, Ji K, Tsirka SE. The experimental autoimmune encephalomyelitis disease course is modulated by nicotine and other cigarette smoke components. *PLoS One* 2014;**9**:e107979.
132. Middlebrook AJ, Martina C, Chang Y, Lukas RJ, DeLuca D. Effects of nicotine exposure on T cell development in fetal thymus organ culture: arrest of T cell maturation. *J Immunol* 2002;**169**:2915–24.
133. Skok MV, Grailhe R, Agenes F, Changeux JP. The role of nicotinic receptors in B-lymphocyte development and activation. *Life Sci* 2007;**80**:2334–6.
134. Nouri-Shirazi M, Tinajero R, Guinet E. Nicotine alters the biological activities of developing mouse bone marrow-derived dendritic cells (DCs). *Immunol Lett* 2007;**109**:155–64.
135. Guinet E, Yoshida K, Nouri-Shirazi M. Nicotinic environment affects the differentiation and functional maturation of monocytes derived dendritic cells (DCs). *Immunol Lett* 2004;**95**:45–55.
136. Baez-Pagan CA, Delgado-Velez M, Lasalde-Dominicci JA. Activation of the macrophage alpha7 nicotinic acetylcholine receptor and control of inflammation. *J Neuroimmune Pharmacol* 2015;**10**:468–76.
137. Kawashima K, Yoshikawa K, Fujii YX, Moriwaki Y, Misawa H. Expression and function of genes encoding cholinergic components in murine immune cells. *Life Sci* 2007;**80**:2314–9.
138. Tracey KJ. Reflex control of immunity. *Nat Rev Immunol* 2009;**9**:418–28.
139. Wang H, Yu M, Ochani M, et al. Nicotinic acetylcholine receptor alpha7 subunit is an essential regulator of inflammation. *Nature* 2003;**421**:384–8.
140. Parrish WR, Rosas-Ballina M, Gallowitsch-Puerta M, et al. Modulation of TNF release by choline requires alpha7 subunit nicotinic acetylcholine receptor-mediated signaling. *Mol Med* 2008;**14**:567–74.
141. Shytle RD, Mori T, Townsend K, et al. Cholinergic modulation of microglial activation by alpha 7 nicotinic receptors. *J Neurochem* 2004;**89**:337–43.
142. Hao J, Simard AR, Turner GH, et al. Attenuation of CNS inflammatory responses by nicotine involves alpha7 and non-alpha7 nicotinic receptors. *Exp Neurol* 2011;**227**:110–9.
143. Simard AR, Gan Y, St-Pierre S, et al. Differential modulation of EAE by alpha9*- and beta2*-nicotinic acetylcholine receptors. *Immunol Cell Biol* 2013;**91**:195–200.
144. Verbitsky M, Rothlin CV, Katz E, Elgoyhen AB. Mixed nicotinic-muscarinic properties of the alpha9 nicotinic cholinergic receptor. *Neuropharmacology* 2000;**39**:2515–24.
145. Jiang W, St-Pierre S, Roy P, Morley BJ, Hao J, Simard AR. Infiltration of CCR2+Ly6Chigh proinflammatory monocytes and neutrophils into the central nervous system is modulated by nicotinic acetylcholine receptors in a model of multiple sclerosis. *J Immunol* 2016;**196**:2095–108.
146. American Psychiatric Association. *Diagnostic and statistical manual of mental disorders (DSM-5®)*. American Psychiatric Pub. Arlington, VA 22209, USA; 2013.
146a. Bienvenu OJ, Davydow DS, Kendler KS. Psychiatric 'diseases' versus behavioral disorders and degree of genetic influence. *Psychol Med* 2011;**41**(1):33–40.

146b. Tocchetto A, Salum GA, Blaya C, Teche S, Isolan L, Bortoluzzi A, et al. Evidence of association between Val66Met polymorphism at BDNF gene and anxiety disorders in a community sample of children and adolescents. *Neurosci Lett* 2011;**502**(3):197–200.
146c. Chen ZY, Jing D, Bath KG, Ieraci A, Khan T, Siao CJ, et al. Genetic variant BDNF (Val-66Met) polymorphism alters anxiety-related behavior. *Science* 2006;**314**(5796):140–3.
146d. Wray NR, James MR, Mah SP, Nelson M, Andrews G, Sullivan PF, Montgomery GW, Birley AJ, Braun A, Martin NG. Anxiety and comorbid measures associated with PLXNA2. *Arch Gen Psychiat* 2007;**64**(3):318–26.
147. van Os J, Kapur S. Schizophrenia. *Lancet* 2009;**374**:635–45.
148. Lakhan SE, Vieira KF. Schizophrenia pathophysiology: are we any closer to a complete model. *Ann General Psychiatry* 2009;**8**:12.
149. Pearlson GD, Folley BS. Schizophrenia, psychiatric genetics, and Darwinian psychiatry: an evolutionary framework. *Schizophr Bull* 2008;**34**:722–33.
150. Wallace TL, Bertrand D. Neuronal alpha7 nicotinic receptors as a target for the treatment of schizophrenia. *Int Rev Neurobiol* 2015;**124**:79–111.
151. Freedman R, Hall M, Adler LE, Leonard S. Evidence in postmortem brain tissue for decreased numbers of hippocampal nicotinic receptors in schizophrenia. *Biol Psychiatry* 1995;**38**:22–33.
152. Mexal S, Berger R, Logel J, Ross RG, Freedman R, Leonard S. Differential regulation of alpha7 nicotinic receptor gene (CHRNA7) expression in schizophrenic smokers. *J Mol Neurosci* 2010;**40**:185–95.
153. Featherstone RE, Siegel SJ. The role of nicotine in schizophrenia. *Int Rev Neurobiol* 2015;**124**:23–78.
154. Dani JA, Bertrand D. Nicotinic acetylcholine receptors and nicotinic cholinergic mechanisms of the central nervous system. *Annu Rev Pharmacol Toxicol* 2007;**47**:699–729.
155. Baumeister AA. The chlorpromazine enigma. *J Hist Neurosci* 2013;**22**:14–29.
156. Brisch R, Saniotis A, Wolf R, et al. The role of dopamine in schizophrenia from a neurobiological and evolutionary perspective: old fashioned, but still in vogue. *Front Psych* 2014;**5**:47.
157. de Leon J, Diaz FJ. A meta-analysis of worldwide studies demonstrates an association between schizophrenia and tobacco smoking behaviors. *Schizophr Res* 2005;**76**:135–57.
158. Steuber TL, Danner F. Adolescent smoking and depression: which comes first? *Addict Behav* 2006;**31**:133–6.
159. Jiang J, See YM, Subramaniam M, Lee J. Investigation of cigarette smoking among male schizophrenia patients. *PLoS One* 2013;**8**:e71343.
160. Kumari V, Postma P. Nicotine use in schizophrenia: the self medication hypotheses. *Neurosci Biobehav Rev* 2005;**29**:1021–34.
161. Keltner NL, Grant JS. Smoke, smoke, smoke that cigarette. *Perspect Psychiatr Care* 2006;**42**:256–61.
162. McCloughen A. The association between schizophrenia and cigarette smoking: a review of the literature and implications for mental health nursing practice. *Int J Ment Health Nurs* 2003;**12**:119–29.
163. Harris JG, Kongs S, Allensworth D, et al. Effects of nicotine on cognitive deficits in schizophrenia. *Neuropsychopharmacology* 2004;**29**:1378–85.
164. Swerdlow NR, Braff DL, Geyer MA. Animal models of deficient sensorimotor gating: what we know, what we think we know, and what we hope to know soon. *Behav Pharmacol* 2000;**11**:185–204.

165. Nespor AA, Tizabi Y. Effects of nicotine on quinpirole-and dizocilpine (MK-801)-induced sensorimotor gating impairments in rats. *Psychopharmacology* 2008;**200**:403–11.
166. Cohen SC, Leckman JF, Bloch MH. Clinical assessment of Tourette syndrome and tic disorders. *Neurosci Biobehav Rev* 2013;**37**:997–1007.
167. Robertson MM, Eapen V, Singer HS, et al. Gilles de la Tourette syndrome. *Nat Rev Dis Primers* 2017;**3**:16097.
168. Jackson GM, Draper A, Dyke K, Pepes SE, Jackson SR. Inhibition, disinhibition, and the control of action in Tourette syndrome. *Trends Cogn Sci* 2015;**19**:655–65.
169. Mink JW. Neurobiology of basal ganglia circuits in Tourette syndrome: faulty inhibition of unwanted motor patterns? *Adv Neurol* 2001;**85**:113–22.
170. Wong DF, Singer HS, Brandt J, et al. D2-like dopamine receptor density in Tourette syndrome measured by PET. *J Nucl Med* 1997;**38**:1243–7.
171. Yoon DY, Gause CD, Leckman JF, Singer HS. Frontal dopaminergic abnormality in Tourette syndrome: a postmortem analysis. *J Neurol Sci* 2007;**255**:50–6.
172. Yeh CB, Lee CH, Chou YH, Chang CJ, Ma KH, Huang WS. Evaluating dopamine transporter activity with 99mTc-TRODAT-1 SPECT in drug-naive Tourette's adults. *Nucl Med Commun* 2006;**27**:779–84.
173. Yeh CB, Lee CS, Ma KH, Lee MS, Chang CJ, Huang WS. Phasic dysfunction of dopamine transmission in Tourette's syndrome evaluated with 99mTc TRODAT-1 imaging. *Psychiatry Res* 2007;**156**:75–82.
174. Abelson JF, Kwan KY, O'Roak BJ, et al. Sequence variants in SLITRK1 are associated with Tourette's syndrome. *Science* 2005;**310**:317–20.
175. Deng H, Le WD, Xie WJ, Jankovic J. Examination of the SLITRK1 gene in Caucasian patients with Tourette syndrome. *Acta Neurol Scand* 2006;**114**:400–2.
176. Leivonen S, Chudal R, Joelsson P, et al. Prenatal maternal smoking and Tourette syndrome: a nationwide register study. *Child Psychiatry Hum Dev* 2016;**47**:75–82.
177. Browne HA, Modabbernia A, Buxbaum JD, et al. Prenatal maternal smoking and increased risk for Tourette syndrome and chronic tic disorders. *J Am Acad Child Adolesc Psychiatry* 2016;**55**:784–91.
178. McConville BJ, Fogelson MH, Norman AB, et al. Nicotine potentiation of haloperidol in reducing tic frequency in Tourette's disorder. *Am J Psychiatry* 1991;**148**:793–4.
179. Sanberg PR, McConville BJ, Fogelson HM, et al. Nicotine potentiates the effects of haloperidol in animals and in patients with Tourette syndrome. *Biomed Pharmacother* 1989;**43**:19–23.
180. Dursun SM, Kutcher S. Smoking, nicotine and psychiatric disorders: evidence for therapeutic role, controversies and implications for future research. *Med Hypotheses* 1999;**52**:101–9.
181. Dursun SM, Reveley MA, Bird R, Stirton F. Longlasting improvement of Tourette's syndrome with transdermal nicotine. *Lancet* 1994;**344**:1577.
182. Silver AA, Shytle RD, Philipp MK, Sanberg PR. Case study: long-term potentiation of neuroleptics with transdermal nicotine in Tourette's syndrome. *J Am Acad Child Adolesc Psychiatry* 1996;**35**:1631–6.
183. Silver AA, Shytle RD, Philipp MK, Wilkinson BJ, McConville B, Sanberg PR. Transdermal nicotine and haloperidol in Tourette's disorder: a double-blind placebo-controlled study. *J Clin Psychiatry* 2001;**62**:707–14.
184. Sanberg PR, Vindrola-Padros C, Shytle RD. Translating laboratory discovery to the clinic: from nicotine and mecamylamine to Tourette's, depression, and beyond. *Physiol Behav* 2012;**107**:801–8.

185. Polanczyk G, de Lima MS, Horta BL, Biederman J, Rohde LA. The worldwide prevalence of ADHD: a systematic review and metaregression analysis. *Am J Psychiatr* 2007;**164**:942–8.
186. Kessler RC, Adler L, Barkley R, et al. The prevalence and correlates of adult ADHD in the United States: results from the National Comorbidity Survey Replication. *Am J Psychiatry* 2006;**163**:716–23.
187. Zhang L, Chang S, Li Z, et al. A genetic database for attention deficit hyperactivity disorder. An overview of ADHDgene. *Nucleic Acid Research* 2012;**40**:D1003–9.
188. Pomerleau OF, Downey KK, Stelson FW, Pomerleau CS. Cigarette smoking in adult patients diagnosed with attention deficit hyperactivity disorder. *J Subst Abus* 1995;**7**:373–8.
189. Milberger S, Biederman J, Faraone SV, Chen L, Jones J. ADHD is associated with early initiation of cigarette smoking in children and adolescents. *J Am Acad Child Adolesc Psychiatry* 1997;**36**:37–44.
190. Lambert NM, Hartsough CS. Prospective study of tobacco smoking and substance dependencies among samples of ADHD and non-ADHD participants. *J Learn Disabil* 1998;**31**:533–44.
191. McClernon FJ, Kollins SH. ADHD and smoking: from genes to brain to behavior. *Ann N Y Acad Sci* 2008;**1141**:131–47.
192. Dougherty DD, Bonab AA, Spencer TJ, Rauch SL, Madras BK, Fischman AJ. Dopamine transporter density in patients with attention deficit hyperactivity disorder. *Lancet* 1999;**354**:2132–3.
193. Fusar-Poli P, Rubia K, Rossi G, Sartori G, Balottin U. Striatal dopamine transporter alterations in ADHD: pathophysiology or adaptation to psychostimulants? A meta-analysis. *Am J Psychiatry* 2012;**169**:264–72.
194. Chandler DJ, Waterhouse BD, Gao WJ. New perspectives on catecholaminergic regulation of executive circuits: evidence for independent modulation of prefrontal functions by midbrain dopaminergic and noradrenergic neurons. *Front Neural Circuits* 2014;**8**:53.
195. Bidwell LC, McClernon FJ, Kollins SH. Cognitive enhancers for the treatment of ADHD. *Pharmacol Biochem Behav* 2011;**99**:262–74.
196. Cortese S. The neurobiology and genetics of attention-deficit/hyperactivity disorder (ADHD): what every clinician should know. *Eur J Paediatr Neurol* 2012;**16**:422–33.
197. Lesch KP, Merker S, Reif A, Novak M. Dances with black widow spiders: dysregulation of glutamate signalling enters centre stage in ADHD. *Eur Neuropsychopharmacol* 2013;**23**:479–91.
198. Giorguieff-Chesselet M, Kemel M, Wandscheer D, Glowinski J. Regulation of dopamine release by presynaptic nicotinic receptors in rat striatal slices: effect of nicotine in a low concentration. *Life Sci* 1979;**25**:1257–61.
199. Brody AL, Olmstead RE, London ED, et al. Smoking-induced ventral striatum dopamine release. *Am J Psychiatry* 2004;**161**:1211–8.
200. Potter AS, Newhouse PA. Acute nicotine improves cognitive deficits in young adults with attention-deficit/hyperactivity disorder. *Pharmacol Biochem Behav* 2008;**88**:407–17.
201. Levin ED, Conners CK, Sparrow E, et al. Nicotine effects on adults with attention-deficit/hyperactivity disorder. *Psychopharmacology (Berlin)* 1996;**123**:55–63.
202. Bayer SA, Altman J, Russo RJ, Zhang X. Timetables of neurogenesis in the human brain based on experimentally determined patterns in the rat. *Neurotoxicology* 1993;**14**:83–144.
203. Quinn R. Comparing rat's to human's age: how old is my rat in people years? *Nutrition* 2005;**21**:775–7.
204. Eppolito AK, Smith RF. Long-term behavioral and developmental consequences of pre- and perinatal nicotine. *Pharmacol Biochem Behav* 2006;**85**:835–41.
205. Peters DA, Taub H, Tang S. Postnatal effects of maternal nicotine exposure. *Neurobehav Toxicol* 1979;**1**:221–5.

206. Richardson SA, Tizabi Y. Hyperactivity in the offspring of nicotine-treated rats: role of the mesolimbic and nigrostriatal dopaminergic pathways. *Pharmacol Biochem Behav* 1994;**47**:331–7.
207. Thomas JD, Garrison ME, Slawecki CJ, Ehlers CL, Riley EP. Nicotine exposure during the neonatal brain growth spurt produces hyperactivity in preweanling rats. *Neurotoxicol Teratol* 2000;**22**:695–701.
208. Vaglenova J, Birru S, Pandiella NM, Breese CR. An assessment of the long-term developmental and behavioral teratogenicity of prenatal nicotine exposure. *Behav Brain Res* 2004;**150**:159–70.
209. LeSage MG, Gustaf E, Dufek MB, Pentel PR. Effects of maternal intravenous nicotine administration on locomotor behavior in pre-weanling rats. *Pharmacol Biochem Behav* 2006;**85**:575–83.
210. Peters DA, Tang S. Sex-dependent biological changes following prenatal nicotine exposure in the rat. *Pharmacol Biochem Behav* 1982;**17**:1077–82.
211. Romero RD, Chen WJ. Gender-related response in open-field activity following developmental nicotine exposure in rats. *Pharmacol Biochem Behav* 2004;**78**:675–81.
212. Ajarem JS, Ahmad M. Prenatal nicotine exposure modifies behavior of mice through early development. *Pharmacol Biochem Behav* 1998;**59**:313–8.
213. Paz R, Barsness B, Martenson T, Tanner D, Allan AM. Behavioral teratogenicity induced by nonforced maternal nicotine consumption. *Neuropsychopharmacology* 2007;**32**:693–9.
214. Pauly JR, Sparks JA, Hauser KF, Pauly TH. In utero nicotine exposure causes persistent, gender-dependant changes in locomotor activity and sensitivity to nicotine in C57Bl/6 mice. *Int J Dev Neurosci* 2004;**22**:329–37.
215. Martin JC, Becker RF. The effects of maternal nicotine absorption or hypoxic episodes upon appetitive behavior of rat offspring. *Dev Psychobiol* 1971;**4**:133–47.
216. Sorenson CA, Raskin LA, Suh Y. The effects of prenatal nicotine on radial-arm maze performance in rats. *Pharmacol Biochem Behav* 1991;**40**:991–3.
217. Yanai J, Pick CG, Rogel-Fuchs Y, Zahalka EA. Alterations in hippocampal cholinergic receptors and hippocampal behaviors after early exposure to nicotine. *Brain Res Bull* 1992;**29**:363–8.
218. Cutler AR, Wilkerson AE, Gingras JL, Levin ED. Prenatal cocaine and/or nicotine exposure in rats: preliminary findings on long-term cognitive outcome and genital development at birth. *Neurotoxicol Teratol* 1996;**18**:635–43.
219. Huang LZ, Liu X, Griffith WH, Winzer-Serhan UH. Chronic neonatal nicotine increases anxiety but does not impair cognition in adult rats. *Behav Neurosci* 2007;**121**:1342–52.
220. Levin ED, Wilkerson A, Jones JP, Christopher NC, Briggs SJ. Prenatal nicotine effects on memory in rats: pharmacological and behavioral challenges. *Brain Res Dev Brain Res* 1996;**97**:207–15.
221. *Depression: a global crisis. World mental health day, October 10 2012*. World Federation; 2012. p. 2012.
222. Lang UE, Borgwardt S. Molecular mechanisms of depression: perspectives on new treatment strategies. *Cell Physiol Biochem* 2013;**31**:761–77.
223. Rohde P, Kahler CW, Lewinsohn PM, Brown RA. Psychiatric disorders, familial factors, and cigarette smoking: II. Associations with progression to daily smoking. *Nicotine Tob Res* 2004;**6**:119–32.
224. Haarasilta LM, Marttunen MJ, Kaprio JA, Aro HM. Correlates of depression in a representative nationwide sample of adolescents (15–19 years) and young adults (20–24 years). *Eur J Pub Health* 2004;**14**:280–5.

225. Hu M-C, Davies M, Kandel DB. Epidemiology and correlates of daily smoking and nicotine dependence among young adults in the United States. *Am J Public Health* 2006;**96**:299–308.
226. Breslau N, Kilbey MM, Andreski P. Nicotine dependence and major depression: new evidence from a prospective investigation. *Arch Gen Psychiatry* 1993;**50**:31–5.
227. Breslau N, Peterson EL, Schultz LR, Chilcoat HD, Andreski P. Major depression and stages of smoking: a longitudinal investigation. *Arch Gen Psychiatry* 1998;**55**:161–6.
228. Klungsøyr O, Nygård JF, Sørensen T, Sandanger I. Cigarette smoking and incidence of first depressive episode: an 11-year, population-based follow-up study. *Am J Epidemiol* 2006;**163**:421–32.
229. Janowsky D, Davis J, El-Yousef MK, Sekerke HJ. A cholinergic-adrenergic hypothesis of mania and depression. *Lancet* 1972;**300**:632–5.
230. Caldarone BJ, Harrist A, Cleary MA, Beech RD, King SL, Picciotto MR. High-affinity nicotinic acetylcholine receptors are required for antidepressant effects of amitriptyline on behavior and hippocampal cell proliferation. *Biol Psychiatry* 2004;**56**:657–64.
231. Brody AL, Mandelkern MA, London ED, et al. Cigarette smoking saturates brain alpha 4 beta 2 nicotinic acetylcholine receptors. *Arch Gen Psychiatry* 2006;**63**:907–15.
232. Staley JK, Krishnan-Sarin S, Cosgrove KP, et al. Human tobacco smokers in early abstinence have higher levels of beta2* nicotinic acetylcholine receptors than nonsmokers. *J Neurosci* 2006;**26**:8707–14.
233. Caldarone BJ, Harrist A, Cleary MA, Beech RD, King SL, Picciotto MR. High-affinity nicotinic acetylcholine receptors are required for antidepressant effects of amitriptyline on behavior and hippocampal cell proliferation. *Biol Psychiatry* 2004;**56**:657–64.
234. Knott V, Thompson A, Shah D, Ilivitsky V. Neural expression of nicotine's antidepressant properties during tryptophan depletion: an EEG study in healthy volunteers at risk for depression. *Biol Psychol* 2012;**91**:190–200.
235. Rabenstein R, Caldarone B, Picciotto M. The nicotinic antagonist mecamylamine has antidepressant-like effects in wild-type but not β2-or α7-nicotinic acetylcholine receptor subunit knockout mice. *Psychopharmacology* 2006;**189**:395–401.
236. Andreasen J, Olsen G, Wiborg O, Redrobe J. Antidepressant-like effects of nicotinic acetylcholine receptor antagonists, but not agonists, in the mouse forced swim and mouse tail suspension tests. *J Psychopharmacol* 2009;**23**:797–804.
237. Barlow DH. Unraveling the mysteries of anxiety and its disorders from the perspective of emotion theory. *Am Psychol* 2000;**55**:1247–63.
238. Vos T, Flaxman AD, Naghavi M, et al. Years lived with disability (YLDs) for 1160 sequelae of 289 diseases and injuries 1990–2010: a systematic analysis for the Global Burden of Disease Study 2010. *Lancet* 2012;**380**:2163–96.
239. Kessler RC, Berglund P, Demler O, Jin R, Merikangas KR, Walters EE. Lifetime prevalence and age-of-onset distributions of DSM-IV disorders in the National Comorbidity Survey Replication. *Arch Gen Psychiatry* 2005;**62**:593–602.
240. Pantelis C, Pantelis C, Yücel M, Wood SJ, McGorry PD, Velakoulis D. Early and late neurodevelopmental disturbances in schizophrenia and their functional consequences. *Aust N Z J Psychiatry* 2003;**37**:399–406.
241. Kessler RC, Adler L, Barkley R, et al. The prevalence and correlates of adult ADHD in the United States: results from the National Comorbidity Survey Replication. *Am J Psychiatr* 2006;**163**:716–23.
242. Yu ZJ, Wecker L. Chronic nicotine administration differentially affects neurotransmitter release from rat striatal slices. *J Neurochem* 1994;**63**:186–94.

243. Salín-Pascual RJ, Rosas M, Jimenez-Genchi A, Rivera-Meza BL. Antidepressant effect of transdermal nicotine patches in nonsmoking patients with major depression. *J Clin Psychiatry* 1996;**57**:387–9.
244. Moylan S, Jacka FN, Pasco JA, Berk M. How cigarette smoking may increase the risk of anxiety symptoms and anxiety disorders: a critical review of biological pathways. *Drug Metab Pharmacokinet* 2013;**3**:302–26.
245. Semba J, Mataki C, Yamada S, Nankai M, Toru M. Antidepressantlike effects of chronic nicotine on learned helplessness paradigm in rats. *Biol Psychiatry* 1998;**43**:389–91.
246. Metcalf RL. *Insect control Ullmann's encyclopedia of industrial chemistry*. 7th ed. Wiley, New York, USA; 2007. p. 9.
247. PubChem. *National Institutes of Health*; 2016.
248. Levin ED, McClernon FJ, Rezvani AH. Nicotinic effects on cognitive function: behavioral characterization, pharmacological specification, and anatomic localization. *Psychopharmacology* 2006;**184**:523–39.
249. Mancuso G, Warburton DM, Mélen M, Sherwood N, Tirelli E. Selective effects of nicotine on attentional processes. *Psychopharmacology* 1999;**146**:199–204.
250. Levin ED, Simon BB. Nicotinic acetylcholine involvement in cognitive function in animals. *Psychopharmacology* 1998;**138**:217–30.
251. Mirza NR, Stolerman IP. Nicotine enhances sustained attention in the rat under specific task conditions. *Psychopharmacology* 1998;**138**:266–74.
252. Puma C, Deschaux O, Molimard R, Bizot J-C. Nicotine improves memory in an object recognition task in rats. *Eur Neuropsychopharmacol* 1999;**9**:323–7.
253. Elrod K, Buccafusco JJ, Jackson WJ. Nicotine enhances delayed matching-to-sample performance by primates. *Life Sci* 1988;**43**:277–87.
254. Buccafusco JJ, Jackson WJ. Beneficial effects of nicotine administered prior to a delayed matching-to-sample task in young and aged monkeys. *Neurobiol Aging* 1991;**12**:233–8.
255. White HK, Levin ED. Four-week nicotine skin patch treatment effects on cognitive performance in Alzheimer's disease. *Psychopharmacology* 1999;**143**:158–65.
256. Newhouse PA, Kelton M. Nicotinic systems in central nervous systems disease: degenerative disorders and beyond. *Pharm Acta Helv* 2000;**74**:91–101.
257. Min S, Moon I-W, Ko R, Shin H. Effects of transdermal nicotine on attention and memory in healthy elderly non-smokers. *Psychopharmacology* 2001;**159**:83–8.
258. Sahakian B, Jones G, Levy R, Gray J, Warburton D. The effects of nicotine on attention, information processing, and short-term memory in patients with dementia of the Alzheimer type. *Br J Psychiatry* 1989;**154**:797–800.
259. Heishman SJ, Kleykamp BA, Singleton EG. Meta-analysis of the acute effects of nicotine and smoking on human performance. *Psychopharmacology (Berlin)* 2010;**210**:453–69.
260. Yakel JL. Gating of nicotinic ACh receptors: latest insights into ligand binding and function. *J Physiol* 2010;**588**:597–602.
261. Gay EA, Yakel JL. Gating of nicotinic ACh receptors; new insights into structural transitions triggered by agonist binding that induce channel opening. *J Physiol* 2007;**584**:727–33.
262. Albuquerque EX, Pereira EF, Alkondon M, Rogers SW. Mammalian nicotinic acetylcholine receptors: from structure to function. *Physiol Rev* 2009;**89**:73–120.
263. Drago J, McColl C, Horne M, Finkelstein D, Ross S. Neuronal nicotinic receptors: insights gained from gene knockout an knocking mutant mice. *Cell Mol Life Sci* 2003;**60**:1267–80.
264. del Toro ED, Juiz JM, Peng X, Lindstrom J, Criado M. Immunocytochemical localization of the α7 subunit of the nicotinic acetylcholine receptor in the rat central nervous system. *J Comp Neurol* 1994;**349**:325–42.

265. Dani JA. Neuronal nicotinic acetylcholine receptor structure and function and response to nicotine. *Nicotine use in mental illness and neurological disorders*. Elsevier, San Diego, United States; 2015.
266. Fucile S. Ca2+ permeability of nicotinic acetylcholine receptors. *Cell Calcium* 2004;**35**:1–8.
267. Fucile S, Sucapane A, Eusebi F. Ca2+ permeability through rat cloned α9-containing nicotinic acetylcholine receptors. *Cell Calcium* 2006;**39**:349–55.
268. Fasoli F, Gotti C. Structure of neuronal nicotinic receptors. *Curr Top Behav Neurosci* 2015;**23**:1–17.
269. Giniatullin R, Nistri A, Yakel JL. Desensitization of nicotinic ACh receptors: shaping cholinergic signaling. *Trends Neurosci* 2005;**28**:371–8.
270. Quick MW, Lester RA. Desensitization of neuronal nicotinic receptors. *J Neurobiol* 2002;**53**:457–78.
271. Singer S, Rossi S, Verzosa S, et al. Nicotine-induced changes in neurotransmitter levels in brain areas associated with cognitive function. *Neurochem Res* 2004;**29**:1779–92.
272. Yasuyoshi H, Kashii S, Kikuchi M, Zhang S, Honda Y, Akaike A. New insight into the functional role of acetylcholine in developing embryonic rat retinal neurons. *Invest Ophthalmol Vis Sci* 2002;**43**:446–51.
273. Markou A. Review. Neurobiology of nicotine dependence. *Philos Trans R Soc Lond B Biol Sci* 2008;**363**:3159–68.
274. Wonnacott S. Presynaptic nicotinic ACh receptors. *Trends Neurosci* 1997;**20**:92–8.
275. Mulle C, Choquet D, Korn H, Changeux J-P. Calcium influx through nicotinic receptor in rat central neurons: its relevance to cellular regulation. *Neuron* 1992;**8**:135–43.
276. Vernino S, Amador M, Luetje CW, Patrick J, Dani JA. Calcium modulation and high calcium permeability of neuronal nicotinic acetylcholine receptors. *Neuron* 1992;**8**:127–34.
277. Rathouz MM, Vijayaraghavan S, Berg DK. Elevation of intracellular calcium levels in neurons by nicotinic acetylcholine receptors. *Mol Neurobiol* 1996;**12**:117–31.
278. Dajas-Bailador FA, Lima PA, Wonnacott S. The α7 nicotinic acetylcholine receptor subtype mediates nicotine protection against NMDA excitotoxicity in primary hippocampal cultures through a Ca2+ dependent mechanism. *Neuropharmacology* 2000;**39**:2799–807.
279. Stevens TR, Krueger SR, Fitzsimonds RM, Picciotto MR. Neuroprotection by nicotine in mouse primary cortical cultures involves activation of calcineurin and L-type calcium channel inactivation. *J Neurosci* 2003;**23**:10093–9.
280. Messi M, Renganathan M, Grigorenko E, Delbono O. Activation of α7 nicotinic acetylcholine receptor promotes survival of spinal cord motoneurons. *FEBS Lett* 1997;**411**:32–8.
281. Ren K, Puig V, Papke RL, Itoh Y, Hughes JA, Meyer EM. Multiple calcium channels and kinases mediate α7 nicotinic receptor neuroprotection in PC12 cells. *J Neurochem* 2005;**94**:926–33.
282. Hosur V, Loring RH. α4β2 nicotinic receptors partially mediate anti-inflammatory effects through Janus kinase 2-signal transducer and activator of transcription 3 but not calcium or cAMP signaling. *Mol Pharmacol* 2011;**79**:167–74.
283. Wang J, Horenstein NA, Stokes C, Papke RL. Tethered agonist analogs as site-specific probes for domains of the human alpha7 nicotinic acetylcholine receptor that differentially regulate activation and desensitization. *Mol Pharmacol* 2010;**78**:1012–25.
284. Grando SA. Connections of nicotine to cancer. *Nat Rev Cancer* 2014;**14**:419–29.
285. Kihara T, Shimohama S, Sawada H, et al. α7 nicotinic receptor transduces signals to phosphatidylinositol 3-kinase to block A β-amyloid-induced neurotoxicity. *J Biol Chem* 2001;**276**:13541–6.
286. Liu Q, Zhao B. Nicotine attenuates β-amyloid peptide-induced neurotoxicity, free radical and calcium accumulation in hippocampal neuronal cultures. *Br J Pharmacol* 2004;**141**:746–54.

287. Onoda N, Nehmi A, Weiner D, Mujumdar S, Christen R, Los G. Nicotine affects the signaling of the death pathway, reducing the response of head and neck cancer cell lines to DNA damaging agents. *Head Neck* 2001;**23**:860–70.
288. Garrido R, Malecki A, Hennig B, Toborek M. Nicotine attenuates arachidonic acid-induced neurotoxicity in cultured spinal cord neurons. *Brain Res* 2000;**861**:59–68.
289. Garrido R, Mattson MP, Hennig B, Toborek M. Nicotine protects against arachidonic-acid-induced caspase activation, cytochrome c release and apoptosis of cultured spinal cord neurons. *J Neurochem* 2001;**76**:1395–403.
290. Dajas-Bailador F, Soliakov L, Wonnacott S. Nicotine activates the extracellular signal-regulated kinase 1/2 via the α7 nicotinic acetylcholine receptor and protein kinase A, in SH-SY5Y cells and hippocampal neurones. *J Neurochem* 2002;**80**:520–30.
291. Shaw S, Bencherif M, Marrero MB. Janus kinase 2, an early target of α7 nicotinic acetylcholine receptor-mediated neuroprotection against Aβ-(1–42) amyloid. *J Biol Chem* 2002;**277**:44920–4.
292. Arredondo J, Chernyavsky AI, Jolkovsky DL, Pinkerton KE, Grando SA. Receptor-mediated tobacco toxicity: cooperation of the Ras/Raf-1/MEK1/ERK and JAK-2/STAT-3 pathways downstream of α7 nicotinic receptor in oral keratinocytes. *FASEB J* 2006;**20**:2093–101.
293. de Jonge WJ, van der Zanden EP, The FO, et al. Stimulation of the vagus nerve attenuates macrophage activation by activating the Jak2-STAT3 signaling pathway. *Nat Immunol* 2005;**6**:844–51.
294. Tracey KJ. Physiology and immunology of the cholinergic antiinflammatory pathway. *J Clin Invest* 2007;**117**:289–96.
295. Hosur V, Leppanen S, Abutaha A, Loring RH. Gene regulation of α4β2 nicotinic receptors: microarray analysis of nicotine-induced receptor up-regulation and anti-inflammatory effects. *J Neurochem* 2009;**111**:848–58.
296. Toulorge D, Guerreiro S, Hild A, Maskos U, Hirsch EC, Michel PP. Neuroprotection of midbrain dopamine neurons by nicotine is gated by cytoplasmic Ca2+. *FASEB J* 2011;**25**:2563–73.
297. Shimohama S. Nicotinic receptor-mediated neuroprotection in neurodegenerative disease models. *Biol Pharm Bull* 2009;**32**:332–6.
298. Dajas-Bailador F, Wonnacott S. Nicotinic acetylcholine receptors and the regulation of neuronal signalling. *Trends Pharmacol Sci* 2004;**25**:317–24.
299. Mudo G, Belluardo N, Fuxe K. Nicotinic receptor agonists as neuroprotective/neurotrophic drugs. Progress in molecular mechanisms. *J Neural Transm* 2007;**114**:135–47.
300. Kawamata J, Shimohama S. Stimulating nicotinic receptors trigger multiple pathways attenuating cytotoxicity in models of Alzheimer's and Parkinson's diseases. *J Alzheimers Dis* 2011;**24**:95–109.
301. Maggio R, Riva M, Vaglini F, et al. Nicotine prevents experimental parkinsonism in rodents and induces striatal increase of neurotrophic factors. *J Neurochem* 1998;**71**:2439–46.
302. Maggio R, Riva M, Vaglini F, Fornai F, Racagni G, Corsini G. Striatal increase of neurotrophic factors as a mechanism of nicotine protection in experimental parkinsonism. *J Neural Transm* 1997;**104**:1113–23.
303. Belluardo N, Blum M, Mudo G, Andbjer B, Fuxe K. Acute intermittent nicotine treatment produces regional increases of basic fibroblast growth factor messenger RNA and protein in the tel- and diencephalon of the rat. *Neuroscience* 1998;**83**:723–40.
304. Belluardo N, Mudò G, Blum M, Itoh N, Agnati L, Fuxe K. Nicotine-induced FGF-2 mRNA in rat brain is preserved during aging. *Neurobiol Aging* 2004;**25**:1333–42.

305. Li XD, Arias E, Jonnala RR, Mruthinti S, Buccafusco JJ. Effect of amyloid peptides on the increase in TrkA receptor expression induced by nicotine in vitro and in vivo. *J Mol Neurosci* 2005;**27**:325–36.
306. Jonnala RR, Terry AV, Buccafusco JJ. Nicotine increases the expression of high affinity nerve growth factor receptors in both in vitro and in vivo. *Life Sci* 2002;**70**:1543–54.
307. French SJ, Humby T, Horner CH, Sofroniew MV, Rattray M. Hippocampal neurotrophin and trk receptor mRNA levels are altered by local administration of nicotine, carbachol and pilocarpine. *Mol Brain Res* 1999;**67**:124–36.
308. Belluardo N, Olsson P, Mudo G, Sommer W, Amato G, Fuxe K. Transcription factor gene expression profiling after acute intermittent nicotine treatment in the rat cerebral cortex. *Neuroscience* 2005;**133**:787–96.
309. Belluardo N, Mudò G, Blum M, Fuxe K. Central nicotinic receptors, neurotrophic factors and neuroprotection. *Behav Brain Res* 2000;**113**:21–34.
310. Massey KA, Zago WM, Berg DK. BDNF up-regulates α7 nicotinic acetylcholine receptor levels on subpopulations of hippocampal interneurons. *Mol Cell Neurosci* 2006;**33**:381–8.
311. Zhou X, Nai Q, Chen M, Dittus JD, Howard MJ, Margiotta JF. Brain-derived neurotrophic factor and trkB signaling in parasympathetic neurons: relevance to regulating α7-containing nicotinic receptors and synaptic function. *J Neurosci* 2004;**24**:4340–50.
312. Formaggio E, Fazzini F, Dalfini A, et al. Nicotine increases the expression of neurotrophin receptor tyrosine kinase receptor A in basal forebrain cholinergic neurons. *Neuroscience* 2010;**166**:580–9.
313. Gergalova G, Lykhmus O, Kalashnyk O, et al. Mitochondria express α7 nicotinic acetylcholine receptors to regulate Ca2+ accumulation and cytochrome c release: study on isolated mitochondria. *PLoS One* 2012;**7**:e31361.
314. Park HJ, Lee PH, Ahn YW, et al. Neuroprotective effect of nicotine on dopaminergic neurons by anti-inflammatory action. *Eur J Neurosci* 2007;**26**:79–89.
315. Rosas-Ballina M, Tracey K. Cholinergic control of inflammation. *J Intern Med* 2009;**265**:663–79.
316. Lin MT, Beal MF. Mitochondrial dysfunction and oxidative stress in neurodegenerative diseases. *Nature* 2006;**443**:787–95.
317. Bindoff L, Birch-Machin M, Cartlidge N, Parker W, Turnbull D. Mitochondrial function in Parkinson's disease. *Lancet* 1989;**334**:49.
318. Mizuno Y, Ohta S, Tanaka M, et al. Deficiencies in complex I subunits of the respiratory chain in Parkinson's disease. *Biochem Biophys Res Commun* 1989;**163**:1450–5.
319. Schapira A, Cooper J, Dexter D, Jenner P, Clark J, Marsden C. Mitochondrial complex I deficiency in Parkinson's disease. *Lancet* 1989;**333**:1269.
320. Manczak M, Anekonda TS, Henson E, Park BS, Quinn J, Reddy PH. Mitochondria are a direct site of A beta accumulation in Alzheimer's disease neurons: implications for free radical generation and oxidative damage in disease progression. *Hum Mol Genet* 2006;**15**:1437–49.
321. Caspersen C, Wang N, Yao J, et al. Mitochondrial Abeta: a potential focal point for neuronal metabolic dysfunction in Alzheimer's disease. *FASEB J* 2005;**19**:2040–1.
322. Cormier A, Morin C, Zini R, Tillement J-P, Lagrue G. Nicotine protects rat brain mitochondria against experimental injuries. *Neuropharmacology* 2003;**44**:642–52.
323. Ferger B, Spratt C, Earl CD, Teismann P, Oertel WH, Kuschinsky K. Effects of nicotine on hydroxyl free radical formation in vitro and on MPTP-induced neurotoxicity in vivo. *Naunyn Schmiedeberg's Arch Pharmacol* 1998;**358**:351–9.
324. Newman MB, Arendash GW, Shytle RD, Bickford PC, Tighe T, Sanberg PR. Nicotine's oxidative and antioxidant properties in CNS. *Life Sci* 2002;**71**:2807–20.

325. Soto-Otero R, Méndez-Álvarez E, Hermida-Ameijeiras Á, López-Real AM, Labandeira-García JL. Effects of (−)-nicotine and (−)-cotinine on 6-hydroxydopamine-induced oxidative stress and neurotoxicity: relevance for Parkinson's disease. *Biochem Pharmacol* 2002;**64**:125–35.
326. Xie Y-X, Bezard E, Zhao B-L. Investigating the receptor-independent neuroprotective mechanisms of nicotine in mitochondria. *J Biol Chem* 2005;**280**:32405–12.
327. Lykhmus O, Gergalova G, Koval L, Zhmak M, Komisarenko S, Skok M. Mitochondria express several nicotinic acetylcholine receptor subtypes to control various pathways of apoptosis induction. *Int J Biochem Cell Biol* 2014;**53**:246–52.
328. Antico Arciuch VG, Alippe Y, Carreras MC, Poderoso JJ. Mitochondrial kinases in cell signaling: facts and perspectives. *Adv Drug Deliv Rev* 2009;**61**:1234–49.
329. Chernyavsky AI, Shchepotin IB, Galitovkiy V, Grando SA. Mechanisms of tumor-promoting activities of nicotine in lung cancer: synergistic effects of cell membrane and mitochondrial nicotinic acetylcholine receptors. *BMC Cancer* 2015;**15**:1.
330. Takarada T, Nakamichi N, Kitajima S, et al. Promoted neuronal differentiation after activation of alpha4/beta2 nicotinic acetylcholine receptors in undifferentiated neural progenitors. *PLoS One* 2012;**7**:e46177.
331. Lee H, Park JR, Yang J, et al. Nicotine inhibits the proliferation by upregulation of nitric oxide and increased HDAC1 in mouse neural stem cells. *In Vitro Cell Dev Biol Anim* 2014;**50**:731–9.
332. He N, Wang Z, Wang Y, Shen H, Yin M. ZY-1, a novel nicotinic analog, promotes proliferation and migration of adult hippocampal neural stem/progenitor cells. *Cell Mol Neurobiol* 2013;**33**:1149–57.
333. Balaraman S, Winzer-Serhan UH, Miranda RC. Opposing actions of ethanol and nicotine on microRNAs are mediated by nicotinic acetylcholine receptors in fetal cerebral cortical-derived neural progenitor cells. *Alcohol Clin Exp Res* 2012;**36**:1669–77.
334. De Simone R, Ajmone-Cat MA, Carnevale D, Minghetti L. Activation of alpha7 nicotinic acetylcholine receptor by nicotine selectively up-regulates cyclooxygenase-2 and prostaglandin E2 in rat microglial cultures. *J Neuroinflammation* 2005;**2**:4.
335. Thomsen MS, Mikkelsen JD. The alpha7 nicotinic acetylcholine receptor ligands methyllycaconitine, NS6740 and GTS-21 reduce lipopolysaccharide-induced TNF-alpha release from microglia. *J Neuroimmunol* 2012;**251**:65–72.
336. Sadigh-Eteghad S, Majdi A, Mahmoudi J, Golzari SE, Talebi M. Astrocytic and microglial nicotinic acetylcholine receptors: an overlooked issue in Alzheimer's disease. *J Neural Transm* 2016;**123**:1359–67.
337. Teaktong T, Graham A, Court J, et al. Alzheimer's disease is associated with a selective increase in alpha7 nicotinic acetylcholine receptor immunoreactivity in astrocytes. *Glia* 2003;**41**:207–11.
338. Parada E, Egea J, Buendia I, et al. The microglial alpha7-acetylcholine nicotinic receptor is a key element in promoting neuroprotection by inducing heme oxygenase-1 via nuclear factor erythroid-2-related factor 2. *Antioxid Redox Signal* 2013;**19**:1135–48.
339. Liu Y, Zeng X, Hui Y, et al. Activation of alpha7 nicotinic acetylcholine receptors protects astrocytes against oxidative stress-induced apoptosis: implications for Parkinson's disease. *Neuropharmacology* 2015;**91**:87–96.
340. Takarada T, Nakamichi N, Kawagoe H, et al. Possible neuroprotective property of nicotinic acetylcholine receptors in association with predominant upregulation of glial cell line-derived neurotrophic factor in astrocytes. *J Neurosci Res* 2012;**90**:2074–85.
341. Konishi Y, Yang LB, He P, et al. Deficiency of GDNF receptor GFRalpha1 in Alzheimer's neurons results in neuronal death. *J Neurosci* 2014;**34**:13127–38.

342. Berridge MS, Apana SM, Nagano KK, Berridge CE, Leisure GP, Boswell MV. Smoking produces rapid rise of [11C]nicotine in human brain. *Psychopharmacology (Berlin)* 2010;**209**:383–94.
343. ChemSpider. *Search and share chemistry*; 2016.
344. Benowitz NL, Hukkanen J, Jacob III P. Nicotine chemistry, metabolism, kinetics and biomarkers. *Nicotine psychopharmacology*. Springer, 2009, p. 29–60.
345. Oldendorf W, Braun L, Cornford E. pH dependence of blood-brain barrier permeability to lactate and nicotine. *Stroke* 1979;**10**:577–81.
346. Fukada A, Saito H, Inui K-I. Transport mechanisms of nicotine across the human intestinal epithelial cell line Caco-2. *J Pharmacol Exp Ther* 2002;**302**:532–8.
347. Fukada A, Saito H, Urakami Y, Okuda M, Inui K-I. Involvement of specific transport system of renal basolateral membranes in distribution of nicotine in rats. *Drug Metab Pharmacokinet* 2002;**17**:554–60.
348. Cisternino S, Chapy H, André P, Smirnova M, Debray M, Scherrmann J-M. Coexistence of passive and proton antiporter-mediated processes in nicotine transport at the mouse blood–brain barrier. *AAPS J* 2013;**15**:299–307.
349. Tega Y, Akanuma S, Kubo Y, Terasaki T, Hosoya K. Blood-to-brain influx transport of nicotine at the rat blood-brain barrier: involvement of a pyrilamine-sensitive organic cation transport process. *Neurochem Int* 2013;**62**:173–81.
350. Tega Y, Akanuma S, Kubo Y, Hosoya K. Involvement of the H+/organic cation antiporter in nicotine transport in rat liver. *Drug Metab Dispos* 2015;**43**:89–92.
351. Takano M, Nagahiro M, Yumoto R. Transport mechanism of nicotine in primary cultured alveolar epithelial cells. *J Pharm Sci* 2016;**105**:982–8.
352. Takami K, Saito H, Okuda M, Takano M, Inui KI. Distinct characteristics of transcellular transport between nicotine and tetraethylammonium in LLC-PK1 cells. *J Pharmacol Exp Ther* 1998;**286**:676–80.
353. Zevin S, Schaner ME, Giacomini KM. Nicotine transport in a human choriocarcinoma cell line (JAR). *J Pharm Sci* 1998;**87**:702–6.
354. Zoli M, Le Novere N, Hill Jr. JA, Changeux JP. Developmental regulation of nicotinic ACh receptor subunit mRNAs in the rat central and peripheral nervous systems. *J Neurosci* 1995;**15**:1912–39.
355. Broide RS, Leslie FM. The alpha7 nicotinic acetylcholine receptor in neuronal plasticity. *Mol Neurobiol* 1999;**20**:1–16.
356. Hellstrom-Lindahl E, Court JA. Nicotinic acetylcholine receptors during prenatal development and brain pathology in human aging. *Behav Brain Res* 2000;**113**:159–68.
357. Hellstrom-Lindahl E, Gorbounova O, Seiger A, Mousavi M, Nordberg A. Regional distribution of nicotinic receptors during prenatal development of human brain and spinal cord. *Brain Res Dev Brain Res* 1998;**108**:147–60.
358. Falk L, Nordberg A, Seiger A, Kjaeldgaard A, Hellstrom-Lindahl E. The alpha7 nicotinic receptors in human fetal brain and spinal cord. *J Neurochem* 2002;**80**:457–65.
359. Fuchs JL. [125I]alpha-Bungarotoxin binding marks primary sensory area developing rat neocortex. *Brain Res* 1989;**501**:223–34.
360. Bina KG, Guzman P, Broide RS, Leslie FM, Smith MA, O'Dowd DK. Localization of alpha 7 nicotinic receptor subunit mRNA and alpha-bungarotoxin binding sites in developing mouse somatosensory thalamocortical system. *J Comp Neurol* 1995;**363**:321–32.
361. Broide RS, O'Connor LT, Smith MA, Smith JA, Leslie FM. Developmental expression of alpha 7 neuronal nicotinic receptor messenger RNA in rat sensory cortex and thalamus. *Neuroscience* 1995;**67**:83–94.

362. Broide RS, Robertson RT, Leslie FM. Regulation of alpha7 nicotinic acetylcholine receptors in the developing rat somatosensory cortex by thalamocortical afferents. *J Neurosci* 1996;**16**:2956–71.
363. Ospina JA, Broide RS, Acevedo D, Robertson RT, Leslie FM. Calcium regulation of agonist binding to alpha7-type nicotinic acetylcholine receptors in adult and fetal rat hippocampus. *J Neurochem* 1998;**70**:1061–8.
364. Adams CE, Broide RS, Chen Y, et al. Development of the alpha7 nicotinic cholinergic receptor in rat hippocampal formation. *Brain Res Dev Brain Res* 2002;**139**:175–87.
365. Agulhon C, Charnay Y, Vallet P, et al. Distribution of mRNA for the alpha4 subunit of the nicotinic acetylcholine receptor in the human fetal brain. *Brain Res Mol Brain Res* 1998;**58**:123–31.
366. Naeff B, Schlumpf M, Lichtensteiger W. Pre- and postnatal development of high-affinity [3H]nicotine binding sites in rat brain regions: an autoradiographic study. *Brain Res Dev Brain Res* 1992;**68**:163–74.
367. Zhang X, Liu C, Miao H, Gong ZH, Nordberg A. Postnatal changes of nicotinic acetylcholine receptor alpha 2, alpha 3, alpha 4, alpha 7 and beta 2 subunits genes expression in rat brain. *Int J Dev Neurosci* 1998;**16**:507–18.
368. Aztiria E, Gotti C, Domenici L. Alpha7 but not alpha4 AChR subunit expression is regulated by light in developing primary visual cortex. *J Comp Neurol* 2004;**480**:378–91.
369. Winzer-Serhan UH, Leslie FM. Codistribution of nicotinic acetylcholine receptor subunit alpha3 and beta4 mRNAs during rat brain development. *J Comp Neurol* 1997;**386**:540–54.
370. Dwyer JB, McQuown SC, Leslie FM. The dynamic effects of nicotine on the developing brain. *Pharmacol Ther* 2009;**122**:125–39.
371. Slotkin TA, Orband-Miller L, Queen KL, Whitmore WL, Seidler FJ. Effects of prenatal nicotine exposure on biochemical development of rat brain regions: maternal drug infusions via osmotic minipumps. *J Pharmacol Exp Ther* 1987;**240**:602–11.
372. Navarro HA, Seidler FJ, Schwartz RD, Baker FE, Dobbins SS, Slotkin TA. Prenatal exposure to nicotine impairs nervous system development at a dose which does not affect viability or growth. *Brain Res Bull* 1989;**23**:187–92.
373. Liedtke W, Edelmann W, Bieri PL, et al. GFAP is necessary for the integrity of CNS white matter architecture and long-term maintenance of myelination. *Neuron* 1996;**17**:607–15.
374. Abdel-Rahman A, Dechkovskaia AM, Sutton JM, et al. Maternal exposure of rats to nicotine via infusion during gestation produces neurobehavioral deficits and elevated expression of glial fibrillary acidic protein in the cerebellum and CA1 subfield in the offspring at puberty. *Toxicology* 2005;**209**:245–61.
375. Roy TS, Sabherwal U. Effects of gestational nicotine exposure on hippocampal morphology. *Neurotoxicol Teratol* 1998;**20**:465–73.
376. Roy TS, Seidler FJ, Slotkin TA. Prenatal nicotine exposure evokes alterations of cell structure in hippocampus and somatosensory cortex. *J Pharmacol Exp Ther* 2002;**300**:124–33.
377. von Ziegler NI, Schlumpf M, Lichtensteiger W. Prenatal nicotine exposure selectively affects perinatal forebrain aromatase activity and fetal adrenal function in male rats. *Brain Res Dev Brain Res* 1991;**62**:23–31.
378. Barbieri RL, Gochberg J, Ryan KJ. Nicotine, cotinine, and anabasine inhibit aromatase in human trophoblast in vitro. *J Clin Invest* 1986;**77**:1727–33.
379. Genedani S, Bernardi M, Bertolini A. Sex-linked differences in avoidance learning in the offspring of rats treated with nicotine during pregnancy. *Psychopharmacology (Berlin)* 1983;**80**:93–5.

380. Liang K, Poytress BS, Chen Y, Leslie FM, Weinberger NM, Metherate R. Neonatal nicotine exposure impairs nicotinic enhancement of central auditory processing and auditory learning in adult rats. *Eur J Neurosci* 2006;**24**:857–66.
381. Paulson RB, Shanfeld J, Vorhees CV, et al. Behavioral effects of prenatally administered smokeless tobacco on rat offspring. *Neurotoxicol Teratol* 1993;**15**:183–92.
382. Staller J, Faraone SV. Attention-deficit hyperactivity disorder in girls: epidemiology and management. *CNS Drugs* 2006;**20**:107–23.
383. Slotkin TA, Orband-Miller L, Queen KL. Development of [3H]nicotine binding sites in brain regions of rats exposed to nicotine prenatally via maternal injections or infusions. *J Pharmacol Exp Ther* 1987;**242**:232–7.
384. Hagino N, Lee JW. Effect of maternal nicotine on the development of sites for [(3)H]nicotine binding in the fetal brain. *Int J Dev Neurosci* 1985;**3**:567–71.
385. Tizabi Y, Perry DC. Prenatal nicotine exposure is associated with an increase in [125I] epibatidine binding in discrete cortical regions in rats. *Pharmacol Biochem Behav* 2000;**67**:319–23.
386. Huang LZ, Winzer-Serhan UH. Chronic neonatal nicotine upregulates heteromeric nicotinic acetylcholine receptor binding without change in subunit mRNA expression. *Brain Res* 2006;**1113**:94–109.
387. Chen K, Nakauchi S, Su H, Tanimoto S, Sumikawa K. Early postnatal nicotine exposure disrupts the alpha2* nicotinic acetylcholine receptor-mediated control of oriens-lacunosum moleculare cells during adolescence in rats. *Neuropharmacology* 2016;**101**:57–67.
388. Tizabi Y, Popke EJ, Rahman MA, Nespor SM, Grunberg NE. Hyperactivity induced by prenatal nicotine exposure is associated with an increase in cortical nicotinic receptors. *Pharmacol Biochem Behav* 1997;**58**:141–6.
389. Muneoka K, Ogawa T, Kamei K, et al. Prenatal nicotine exposure affects the development of the central serotonergic system as well as the dopaminergic system in rat offspring: involvement of route of drug administrations. *Brain Res Dev Brain Res* 1997;**102**:117–26.
390. Arnsten AF. Stimulants: therapeutic actions in ADHD. *Neuropsychopharmacology* 2006;**31**:2376–83.
391. Ribary U, Lichtensteiger W. Effects of acute and chronic prenatal nicotine treatment on central catecholamine systems of male and female rat fetuses and offspring. *J Pharmacol Exp Ther* 1989;**248**:786–92.
392. Oades RD. Role of the serotonin system in ADHD: treatment implications. *Expert Rev Neurother* 2007;**7**:1357–74.
393. Zimmermann GR, Lehar J, Keith CT. Multi-target therapeutics: when the whole is greater than the sum of the parts. *Drug Discov Today* 2007;**12**:34–42.
394. Morphy R, Kay C, Rankovic Z. From magic bullets to designed multiple ligands. *Drug Discov Today* 2004;**9**:641–51.
395. Youdim MB, Buccafusco JJ. Multi-functional drugs for various CNS targets in the treatment of neurodegenerative disorders. *Trends Pharmacol Sci* 2005;**26**:27–35.
396. Shih J, Chen K, Ridd M. Monoamine oxidase: from genes to behavior. *Annu Rev Neurosci* 1999;**22**:197.
397. Riederer P, Lachenmayer L, Laux G. Clinical applications of MAO-inhibitors. *Curr Med Chem* 2004;**11**:2033–43.
398. Youdim MB, Bakhle Y. Monoamine oxidase: isoforms and inhibitors in Parkinson's disease and depressive illness. *Br J Pharmacol* 2006;**147**:S287–96.
399. Berlin I, Anthenelli RM. Monoamine oxidases and tobacco smoking. *Int J Neuropsychopharmacol* 2001;**4**:33–42.

400. Berlin I, Said S, Spreux-Varoquaux O, Olivares R, Launay JM, Puech AJ. Monoamine oxidase A and B activities in heavy smokers. *Biol Psychiatry* 1995;**38**:756–61.
401. Coursey RD, Buchsbaum MS, Murphy DL. Platelet MAO activity and evoked potentials in the identification of subjects biologically at risk for psychiatric disorders. *Br J Psychiatry* 1979;**134**:372–81.
402. Oreland L, Fowler CJ, Schalling D. Low platelet monoamine oxidase activity in cigarette smokers. *Life Sci* 1981;**29**:2511–8.
403. Fowler JS, Volkow ND, Wang GJ, et al. Inhibition of monoamine oxidase B in the brains of smokers. *Nature* 1996;**379**:733–6.
404. Fowler JS, Volkow ND, Wang GJ, et al. Brain monoamine oxidase A inhibition in cigarette smokers. *Proc Natl Acad Sci U S A* 1996;**93**:14065–9.
405. Fowler JS, Volkow ND, Wang GJ, et al. Neuropharmacological actions of cigarette smoke: brain monoamine oxidase B (MAO B) inhibition. *J Addict Dis* 1998;**17**:23–34.
406. Fowler JS, Logan J, Wang GJ, Volkow ND. Monoamine oxidase and cigarette smoking. *Neurotoxicology* 2003;**24**:75–82.
407. Fowler JS, Logan J, Wang GJ, et al. Low monoamine oxidase B in peripheral organs in smokers. *Proc Natl Acad Sci U S A* 2003;**100**:11600–5.
408. Fowler JS, Logan J, Wang GJ, et al. Monoamine oxidase A imaging in peripheral organs in healthy human subjects. *Synapse* 2003;**49**:178–87.
409. Fowler J, Volkow N, Wang G-J, et al. Age-related increases in brain monoamine oxidase B in living healthy human subjects. *Neurobiol Aging* 1997;**18**:431–5.
410. Danielczyk W, Streifler M, Konradi C, Riederer P, Moll G. Platelet MAO-B activity and the psychopathology of Parkinson's disease, senile dementia and multi-infarct dementia. *Acta Psychiatr Scand* 1988;**78**:730–6.
411. Zhou G, Miura Y, Shoji H, Yamada S, Matsuishi T. Platelet monoamine oxidase B and plasma β-phenylethylamine in Parkinson's disease. *J Neurol Neurosurg Psychiatry* 2001;**70**:229–31.
412. Emilsson L, Saetre P, Balciuniene J, Castensson A, Cairns N, Jazin EE. Increased monoamine oxidase messenger RNA expression levels in frontal cortex of Alzheimer's disease patients. *Neurosci Lett* 2002;**326**:56–60.
413. Takehashi M, Tanaka S, Masliah E, Ueda K. Association of monoamine oxidase A gene polymorphism with Alzheimer's disease and Lewy body variant. *Neurosci Lett* 2002;**327**:79–82.
414. Wu YH, Feenstra MG, Zhou JN, et al. Molecular changes underlying reduced pineal melatonin levels in Alzheimer disease: alterations in preclinical and clinical stages. *J Clin Endocrinol Metab* 2003;**88**:5898–906.
415. Yu P, Boulton A. Irreversible inhibition of monoamine oxidase by some components of cigarette smoke. *Life Sci* 1987;**41**:675–82.
416. Essman WB. Serotonin and monoamine oxidase in mouse skin: effects of cigarette smoke exposure. *J Med* 1976;**8**:95–101.
417. Castagnoli K, Steyn SJ, Magnin G, et al. Studies on the interactions of tobacco leaf and tobacco smoke constituents and monoamine oxidase. *Neurotox Res* 2002;**4**:151–60.
418. Costello MR, Reynaga DD, Mojica CY, Zaveri NT, Belluzzi JD, Leslie FM. Comparison of the reinforcing properties of nicotine and cigarette smoke extract in rats. *Neuropsychopharmacology* 2014;**39**:1843–51.
419. Herraiz T, Chaparro C. Human monoamine oxidase is inhibited by tobacco smoke: β-carboline alkaloids act as potent and reversible inhibitors. *Biochem Biophys Res Commun* 2005;**326**:378–86.

420. Khalil AA, Steyn S, Castagnoli N. Isolation and characterization of a monoamine oxidase inhibitor from tobacco leaves. *Chem Res Toxicol* 2000;**13**:31–5.
421. Hauptmann N, Shih JC. 2-Naphthylamine, a compound found in cigarette smoke, decreases both monoamine oxidase A and B catalytic activity. *Life Sci* 2001;**68**:1231–41.
422. Méndez-Álvarez E, Soto-Otero R, Sánchez-Sellero I, Lamas ML-R. Inhibition of brain monoamine oxidase by adducts of 1, 2, 3, 4-tetrahydroisoquinoline with components of cigarette smoke. *Life Sci* 1997;**60**:1719–27.
423. Muriel P, Pérez-Rojas JM. Nitric oxide inhibits mitochondrial monoamine oxidase activity and decreases outer mitochondria membrane fluidity. *Comp Biochem Physiol C Toxicol Pharmacol* 2003;**136**:191–7.
424. Singh T, Williams K. Atypical depression. *Psychiatry* 2006;**3**:33–9.
425. Sathyanarayana Rao TS, Yeragani VK. Hypertensive crisis and cheese. *Indian J Psychiatry* 2009;**51**:65–6.
426. Knoll J, Magyar K. Some puzzling pharmacological effects of monoamine oxidase inhibitors. *Adv Biochem Psychopharmacol* 1972;**5**:393–408.
427. Varga E, Tringer L. Clinical trial of a new type promptly acting psychoenergetic agent (phenyl-isopropyl-methylpropinyl-HCl, "E-250"). *Acta Med Acad Sci Hung* 1967;**23**:289–95.
428. Birkmayer W, Knoll J, Riederer P, Youdim MB, Hars V, Marton J. Increased life expectancy resulting from addition of L-deprenyl to Madopar treatment in Parkinson's disease: a long-term study. *J Neural Transm* 1985;**64**:113–27.
429. Magyar K, Szende B. (−)-Deprenyl, a selective MAO-B inhibitor, with apoptotic and anti-apoptotic properties. *Neurotoxicology* 2004;**25**:233–42.
430. Ebadi M, Brown-Borg H, Ren J, et al. Therapeutic efficacy of selegiline in neurodegenerative disorders and neurological diseases. *Curr Drug Targets* 2006;**7**:1513–29.
431. Thomas CE, Huber EW, Ohlweiler DF. Hydroxyl and peroxyl radical trapping by the monoamine oxidase-B inhibitors deprenyl and MDL 72,974A: implications for protection of biological substrates. *Free Radic Biol Med* 1997;**22**:733–7.
432. Carrillo MC, Kanai S, Nokubo M, Kitani K. (−) Deprenyl induces activities of both superoxide dismutase and catalase but not of glutathione peroxidase in the striatum of young male rats. *Life Sci* 1991;**48**:517–21.
433. Andoh T, Chock PB, Murphy DL, Chiueh CC. Role of the redox protein thioredoxin in cytoprotective mechanism evoked by (−)-deprenyl. *Mol Pharmacol* 2005;**68**:1408–14.
434. Knoll J. The striatal dopamine dependency of life span in male rats. Longevity study with (−) deprenyl. *Mech Ageing Dev* 1988;**46**:237–62.
435. Kitani K, Kanai S, Sato Y, Ohta M, Ivy GO, Carrillo MC. Chronic treatment of (−)deprenyl prolongs the life span of male Fischer 344 rats. Further evidence. *Life Sci* 1993;**52**:281–8.
436. Magyar K, Szende B, Lengyel J, Tekes K. The pharmacology of B-type selective monoamine oxidase inhibitors; milestones in (−)-deprenyl research. *J Neural Transm Suppl* 1996;**48**:29–43.
437. Gotz ME, Breithaupt W, Sautter J, et al. Chronic TVP-1012 (rasagiline) dose–activity response of monoamine oxidases A and B in the brain of the common marmoset. *J Neural Transm Suppl* 1998;**52**:271–8.
438. Heikkila RE, Duvoisin RC, Finberg JP, Youdim MB. Prevention of MPTP-induced neurotoxicity by AGN-1133 and AGN-1135, selective inhibitors of monoamine oxidase-B. *Eur J Pharmacol* 1985;**116**:313–7.
439. Spooren WP, Waldmeier P, Gentsch C. The effect of a subchronic post-lesion treatment with (−)-deprenyl on the sensitivity of 6-OHDA-lesioned rats to apomorphine and d-amphetamine. *J Neural Transm* 1999;**106**:825–33.

440. Kupsch A, Sautter J, Gotz ME, et al. Monoamine oxidase-inhibition and MPTP-induced neurotoxicity in the non-human primate: comparison of rasagiline (TVP 1012) with selegiline. *J Neural Transm* 2001;**108**:985–1009.
441. Finnegan KT, Skratt JJ, Irwin I, DeLanney LE, Langston JW. Protection against DSP-4-induced neurotoxicity by deprenyl is not related to its inhibition of MAO B. *Eur J Pharmacol* 1990;**184**:119–26.
442. Yu PH, Davis BA, Fang J, Boulton AA. Neuroprotective effects of some monoamine oxidase-B inhibitors against DSP-4-induced noradrenaline depletion in the mouse hippocampus. *J Neurochem* 1994;**63**:1820–8.
443. Ricci A, Mancini M, Strocchi P, Bongrani S, Bronzetti E. Deficits in cholinergic neurotransmission markers induced by ethylcholine mustard aziridinium (AF64A) in the rat hippocampus: sensitivity to treatment with the monoamine oxidase-B inhibitor L-deprenyl. *Drugs Exp Clin Res* 1992;**18**:163–71.
444. Speiser Z, Katzir O, Rehavi M, Zabarski T, Cohen S. Sparing by rasagiline (TVP-1012) of cholinergic functions and behavior in the postnatal anoxia rat. *Pharmacol Biochem Behav* 1998;**60**:387–93.
445. Gelowitz DL, Paterson IA. Neuronal sparing and behavioral effects of the antiapoptotic drug, (−)deprenyl, following kainic acid administration. *Pharmacol Biochem Behav* 1999;**62**:255–62.
446. Huang W, Chen Y, Shohami E, Weinstock M. Neuroprotective effect of rasagiline, a selective monoamine oxidase-B inhibitor, against closed head injury in the mouse. *Eur J Pharmacol* 1999;**366**:127–35.
447. Stefanova N, Poewe W, Wenning GK. Rasagiline is neuroprotective in a transgenic model of multiple system atrophy. *Exp Neurol* 2008;**210**:421–7.
448. Maruyama W, Takahashi T, Naoi M. (−)-Deprenyl protects human dopaminergic neuroblastoma SH-SY5Y cells from apoptosis induced by peroxynitrite and nitric oxide. *J Neurochem* 1998;**70**:2510–5.
449. Naoi M, Maruyama W. Functional mechanism of neuroprotection by inhibitors of type B monoamine oxidase in Parkinson's disease. *Expert Rev Neurother* 2009;**9**:1233–50.
450. Tatton WG, Ju WY, Holland DP, Tai C, Kwan M. (−)-Deprenyl reduces PC12 cell apoptosis by inducing new protein synthesis. *J Neurochem* 1994;**63**:1572–5.
451. Semkova I, Wolz P, Schilling M, Krieglstein J. Selegiline enhances NGF synthesis and protects central nervous system neurons from excitotoxic and ischemic damage. *Eur J Pharmacol* 1996;**315**:19–30.
452. Finberg JP, Takeshima T, Johnston JM, Commissiong JW. Increased survival of dopaminergic neurons by rasagiline, a monoamine oxidase B inhibitor. *Neuroreport* 1998;**9**:703–7.
453. Mytilineou C, Leonardi EK, Radcliffe P, et al. Deprenyl and desmethylselegiline protect mesencephalic neurons from toxicity induced by glutathione depletion. *J Pharmacol Exp Ther* 1998;**284**:700–6.
454. Maruyama W, Takahashi T, Youdim M, Naoi M. The anti-Parkinson drug, rasagiline, prevents apoptotic DNA damage induced by peroxynitrite in human dopaminergic neuroblastoma SH-SY5Y cells. *J Neural Transm* 2002;**109**:467–81.
455. Naoi M, Maruyama W, Akao Y, Zhang J, Parvez H. Apoptosis induced by an endogenous neurotoxin, N-methyl(R)salsolinol, in dopamine neurons. *Toxicology* 2000;**153**:123–41.
456. Wu Y, Kazumura K, Maruyama W, Osawa T, Naoi M. Rasagiline and selegiline suppress calcium efflux from mitochondria by PK11195-induced opening of mitochondrial permeability transition pore: a novel anti-apoptotic function for neuroprotection. *J Neural Transm* 2015;**122**:1399–407.

457. Naoi M, Maruyama W. Monoamine oxidase inhibitors as neuroprotective agents in age-dependent neurodegenerative disorders. *Curr Pharm Des* 2010;**16**:2799–817.
458. Gallagher DA, Schrag A. Impact of newer pharmacological treatments on quality of life in patients with Parkinson's disease. *CNS Drugs* 2008;**22**:563–86.
459. Daly JW. Nicotinic agonists, antagonists, and modulators from natural sources. *Cell Mol Neurobiol* 2005;**25**:513–52.
460. Maciuk A, Moaddel R, Haginaka J, Wainer IW. Screening of tobacco smoke condensate for nicotinic acetylcholine receptor ligands using cellular membrane affinity chromatography columns and missing peak chromatography. *J Pharm Biomed Anal* 2008;**48**:238–46.
461. Rodgman A, Perfetti TA. The chemical components identified in tobacco and tobacco smoke prior to 1954: a chronology of classical chemistry. *Beitr Tabakforschung/Contrib Tob Res* 2009;**23**:277–333.
462. Paris D, Beaulieu-Abdelahad D, Bachmeier C, et al. Anatabine lowers Alzheimer's Aβ production in vitro and in vivo. *Eur J Pharmacol* 2011;**670**:384–91.
463. Verma M, Beaulieu-Abdelahad D, Ait-Ghezala G, et al. Chronic anatabine treatment reduces Alzheimer's disease (AD)-like pathology and improves socio-behavioral deficits in a transgenic mouse model of AD. *PLoS One* 2015;**10**:e0128224.
464. Paris D, Beaulieu-Abdelahad D, Ait-Ghezala G, Mathura V, Verma M, Roher A, Reed J, Crawford F, Mullan M. Anatabine attenuates Tau phosphorylation and oligomerization in P301S Tau transgenic mice. *Brain Disord. Ther.* 2014;**3**:126.
465. Paris D, Beaulieu-Abdelahad D, Mullan M, et al. Amelioration of experimental autoimmune encephalomyelitis by anatabine. *PLoS One* 2013;**8**:e55392.
466. Hunter BE, Christopher M, Papke RL, Kem WR, Meyer EM. A novel nicotinic agonist facilitates induction of long-term potentiation in the rat hippocampus. *Neurosci Lett* 1994;**168**:130–4.
467. Levin ED, Hao I, Burke DA, Cauley M, Hall BJ, Rezvani AH. Effects of tobacco smoke constituents, anabasine and anatabine, on memory and attention in female rats. *J Psychopharmacol* 2014;**28**:915–22. https://doi.org/10.1177/0269881114543721.
468. Dwoskin LP, Teng L, Buxton ST, Ravard A, Deo N, Crooks PA. Minor alkaloids of tobacco release [3H]dopamine from superfused rat striatal slices. *Eur J Pharmacol* 1995;**276**:195–9.
469. Zhao F, Gao Z, Jiao W, Chen L, Yao X. In vitro anti-inflammatory effects of beta-carboline alkaloids, isolated from Picrasma quassioides, through inhibition of the iNOS pathway. *Planta Med* 2012;**78**:1906–11.
470. Jiao WH, Gao H, Zhao F, et al. Anti-inflammatory alkaloids from the stems of Picrasma quassioides BENNET. *Chem Pharm Bull (Tokyo)* 2011;**59**:359–64.
471. Gruss M, Appenroth D, Flubacher A, et al. 9-Methyl-beta-carboline-induced cognitive enhancement is associated with elevated hippocampal dopamine levels and dendritic and synaptic proliferation. *J Neurochem* 2012;**121**:924–31.
472. Kallianos A, Shelburne F, Means R, Stevens R, Lax R, Mold J. Identification of the D-glucosides of stigmasterol, sitosterol and campesterol in tobacco and cigarette smoke. *Biochem J* 1963;**87**:596.
473. Shi C, Wu F, Xu J. Incorporation of beta-sitosterol into mitochondrial membrane enhances mitochondrial function by promoting inner mitochondrial membrane fluidity. *J Bioenerg Biomembr* 2013;**45**:301–5.
474. Santos SA, Freire CS, Domingues MR, Silvestre AJ, Pascoal Neto C. Characterization of phenolic components in polar extracts of Eucalyptus globulus Labill. bark by high-performance liquid chromatography-mass spectrometry. *J Agric Food Chem* 2011;**59**:9386–93.

475. Choudhary MI, Naheed N, Abbaskhan A, Musharraf SG, Siddiqui H, Atta UR. Phenolic and other constituents of fresh water fern Salvinia molesta. *Phytochemistry* 2008;**69**:1018–23.
476. Lee YS, Kang YH, Jung JY, et al. Protein glycation inhibitors from the fruiting body of Phellinus linteus. *Biol Pharm Bull* 2008;**31**:1968–72.
477. Rodgman A, Perfetti TA. *The chemical components of tobacco and tobacco smoke*. CRC Press, Boca Raton, Florida, USA; 2013.
478. Ye Z, Liu Z, Henderson A, et al. Increased CYP4B1 mRNA is associated with the inhibition of dextran sulfate sodium-induced colitis by caffeic acid in mice. *Exp Biol Med* 2009;**234**:605–16.
479. Larrosa M, Luceri C, Vivoli E, et al. Polyphenol metabolites from colonic microbiota exert anti-inflammatory activity on different inflammation models. *Mol Nutr Food Res* 2009;**53**:1044–54.
480. Kim JH, Wang Q, Choi JM, Lee S, Cho EJ. Protective role of caffeic acid in an Abeta25-35-induced Alzheimer's disease model. *Nutr Res Pract* 2015;**9**:480–8.
481. Fontanilla CV, Ma Z, Wei X, et al. Caffeic acid phenethyl ester prevents 1-methyl-4-phenyl-1,2,3,6-tetrahydropyridine-induced neurodegeneration. *Neuroscience* 2011;**188**:135–41.
482. Ferchmin PA, Pagan OR, Ulrich H, Szeto AC, Hann RM, Eterovic VA. Actions of octocoral and tobacco cembranoids on nicotinic receptors. *Toxicon* 2009;**54**:1174–82.
483. Ferchmin PA, Hao J, Perez D, et al. Tobacco cembranoids protect the function of acute hippocampal slices against NMDA by a mechanism mediated by alpha4beta2 nicotinic receptors. *J Neurosci Res* 2005;**82**:631–41.
484. Eterovic VA, Perez D, Martins AH, Cuadrado BL, Carrasco M, Ferchmin PA. A cembranoid protects acute hippocampal slices against paraoxon neurotoxicity. *Toxicol In Vitro* 2011;**25**:1468–74.
485. Hann RM, Pagán OR, Gregory L, et al. Characterization of cembranoid interaction with the nicotinic acetylcholine receptor. *J Pharmacol Exp Ther* 1998;**287**:253–60.
486. Eterovic VA, Del Valle-Rodriguez A, Perez D, et al. Protective activity of (1S,2E,4R,6R,7E,11E)-2,7,11-cembratriene-4,6-diol analogues against diisopropylfluorophosphate neurotoxicity: preliminary structure-activity relationship and pharmacophore modeling. *Bioorg Med Chem* 2013;**21**:4678–86.
487. Ferchmin PA, Perez D, Cuadrado BL, Carrasco M, Martins AH, Eterovic VA. Neuroprotection against diisopropylfluorophosphate in acute hippocampal slices. *Neurochem Res* 2015;**40**:2143–51.
488. Olsson E, Holth A, Kumlin E, Bohlin L, Wahlberg I. Structure-related inhibiting activity of some tobacco cembranoids on the prostaglandin synthesis in vitro. *Planta Med* 1993;**59**:293–5.
489. Clifford MN. Chlorogenic acids and other cinnamates – nature, occurrence and dietary burden. *J Sci Food Agric* 1999;**79**:362–72.
490. Zucker M, Ahrens JF. Quantitative assay of chlorogenic acid and its pattern of distribution within tobacco leaves. *Plant Physiol* 1958;**33**:246.
491. Shen W, Qi R, Zhang J, et al. Chlorogenic acid inhibits LPS-induced microglial activation and improves survival of dopaminergic neurons. *Brain Res Bull* 2012;**88**:487–94.
492. Heitman E, Ingram DK. Cognitive and neuroprotective effects of chlorogenic acid. *Nutr Neurosci* 2017;**20**:32–9.
493. Kim J, Lee S, Shim J, et al. Caffeinated coffee, decaffeinated coffee, and the phenolic phytochemical chlorogenic acid up-regulate NQO1 expression and prevent H(2)O(2)-induced apoptosis in primary cortical neurons. *Neurochem Int* 2012;**60**:466–74.

494. Kwon SH, Lee HK, Kim JA, et al. Neuroprotective effects of chlorogenic acid on scopolamine-induced amnesia via anti-acetylcholinesterase and anti-oxidative activities in mice. *Eur J Pharmacol* 2010;**649**:210–7.
495. Gao J, Adam B-L, Terry AV. Evaluation of nicotine and cotinine analogs as potential neuroprotective agents for Alzheimer's disease. *Bioorg Med Chem Lett* 2014;**24**:1472–8.
496. Terry AV, Hernandez CM, Hohnadel EJ, Bouchard KP, Buccafusco JJ. Cotinine, a neuroactive metabolite of nicotine: potential for treating disorders of impaired cognition. *CNS Drug Rev* 2005;**11**:229–52.
497. Burgess S, Zeitlin R, Echeverria V. Cotinine inhibits amyloid-β peptide neurotoxicity and oligomerization. *J.Clin. Toxicol* 2012;S6.
498. Patel S, Grizzell JA, Holmes R, et al. Cotinine halts the advance of Alzheimer's disease-like pathology and associated depressive-like behavior in Tg6799 mice. *Front Aging Neurosci* 2014;**6**:162.
499. Vainio PJ, Tuominen RK. Cotinine binding to nicotinic acetylcholine receptors in bovine chromaffin cell and rat brain membranes. *Nicotine Tob Res* 2001;**3**:177–82.
500. O'Leary K, Parameswaran N, McIntosh JM, Quik M. Cotinine selectively activates a subpopulation of alpha3/alpha6beta2 nicotinic receptors in monkey striatum. *J Pharmacol Exp Ther* 2008;**325**:646–54.
501. Echeverria V, Zeitlin R, Burgess S, et al. Cotinine reduces amyloid-beta aggregation and improves memory in Alzheimer's disease mice. *J Alzheimers Dis* 2011;**24**:817–35.
502. Grizzell JA, Mullins M, Iarkov A, Rohani A, Charry LC, Echeverria V. Cotinine reduces depressive-like behavior and hippocampal vascular endothelial growth factor downregulation after forced swim stress in mice. *Behav Neurosci* 2014;**128**:713.
503. Hatsukami DK, Grillo M, Pentel PR, Oncken C, Bliss R. Safety of cotinine in humans: physiologic, subjective, and cognitive effects. *Pharmacol Biochem Behav* 1997;**57**:643–50.
504. Benowitz NL. Pharmacology of nicotine: addiction and therapeutics. *Annu Rev Pharmacol Toxicol* 1996;**36**:597–613.
505. Rosecrans JA. Nicotine as a discriminative stimulus to behavior: its characterization and relevance to smoking behavior. *NIDA Res Monogr* 1979;**23**:58–69.
506. Fernandez MA, Saenz MT, Garcia MD. Anti-inflammatory activity in rats and mice of phenolic acids isolated from Scrophularia frutescens. *J Pharm Pharmacol* 1998;**50**:1183–6.
507. Yan JJ, Cho JY, Kim HS, et al. Protection against beta-amyloid peptide toxicity in vivo with long-term administration of ferulic acid. *Br J Pharmacol* 2001;**133**:89–96.
508. Wenk GL, McGann-Gramling K, Hauss-Wegrzyniak B, et al. Attenuation of chronic neuroinflammation by a nitric oxide-releasing derivative of the antioxidant ferulic acid. *J Neurochem* 2004;**89**:484–93.
509. Adams JD, LaVoie EJ, Shigematsu A, Owens P, Hoffmann D. Quinoline and methylquinolines in cigarette smoke: comparative data and the effect of filtration. *J Anal Toxicol* 1983;**7**:293–6.
510. Foley M, Tilley L. Quinoline antimalarials: mechanisms of action and resistance and prospects for new agents. *Pharmacol Ther* 1998;**79**:55–87.
511. Fiorito J, Saeed F, Zhang H, et al. Synthesis of quinoline derivatives: discovery of a potent and selective phosphodiesterase 5 inhibitor for the treatment of Alzheimer's disease. *Eur J Med Chem* 2013;**60**:285–94.
512. Adlard PA, Cherny RA, Finkelstein DI, et al. Rapid restoration of cognition in Alzheimer's transgenic mice with 8-hydroxy quinoline analogs is associated with decreased interstitial Abeta. *Neuron* 2008;**59**:43–55.
513. Dressler H. *Resorcinol: its uses and derivatives*. Springer, New York, USA; 1994.

514. Forbes E, Murase T, Yang M, et al. Immunopathogenesis of experimental ulcerative colitis is mediated by eosinophil peroxidase. *J Immunol* 2004;**172**:5664–75.
515. Granja AG, Carrillo-Salinas F, Pagani A, et al. A cannabigerol quinone alleviates neuroinflammation in a chronic model of multiple sclerosis. *J Neuroimmune Pharmacol* 2012;**7**:1002–16.
516. Xu P-x, Wang S-w, Yu X-l, et al. Rutin improves spatial memory in Alzheimer's disease transgenic mice by reducing Aβ oligomer level and attenuating oxidative stress and neuroinflammation. *Behav Brain Res* 2014;**264**:173–80.
517. Pyrzanowska J, Piechal A, Blecharz-Klin K, Joniec-Maciejak I, Zobel A, Widy-Tyszkiewicz E. Influence of long-term administration of rutin on spatial memory as well as the concentration of brain neurotransmitters in aged rats. *Pharmacol Rep* 2012;**64**:808–16.
518. Nassiri-Asl M, Mortazavi SR, Samiee-Rad F, et al. The effects of rutin on the development of pentylenetetrazole kindling and memory retrieval in rats. *Epilepsy Behav* 2010;**18**:50–3.
519. Koda T, Kuroda Y, Imai H. Protective effect of rutin against spatial memory impairment induced by trimethyltin in rats. *Nutr Res* 2008;**28**:629–34.
520. Khan MM, Raza SS, Javed H, et al. Rutin protects dopaminergic neurons from oxidative stress in an animal model of Parkinson's disease. *Neurotox Res* 2012;**22**:1–15.
521. Magalingam KB, Radhakrishnan A, Haleagrahara N. Rutin, a bioflavonoid antioxidant protects rat pheochromocytoma (PC-12) cells against 6-hydroxydopamine (6-OHDA)-induced neurotoxicity. *Int J Mol Med* 2013;**32**:235–40.
522. Jimenez-Aliaga K, Bermejo-Bescos P, Benedi J, Martin-Aragon S. Quercetin and rutin exhibit antiamyloidogenic and fibril-disaggregating effects in vitro and potent antioxidant activity in APPswe cells. *Life Sci* 2011;**89**:939–45.
523. Wang SW, Wang YJ, Su YJ, et al. Rutin inhibits beta-amyloid aggregation and cytotoxicity, attenuates oxidative stress, and decreases the production of nitric oxide and proinflammatory cytokines. *Neurotoxicology* 2012;**33**:482–90.
524. Yang C-H, Nakagawa Y, Wender SH. Identification of scopoletin in cigarette tobacco and smoke. *J Org Chem* 1958;**23**:204–5.
525. Calixto JB, Otuki MF, Santos AR. Anti-inflammatory compounds of plant origin. Part I. Action on arachidonic acid pathway, nitric oxide and nuclear factor kappa B (NF-kappaB). *Planta Med* 2003;**69**:973–83.
526. Muschietti L, Gorzalczany S, Ferraro G, Acevedo C, Martino V. Phenolic compounds with anti-inflammatory activity from Eupatorium buniifolium. *Planta Med* 2001;**67**:743–4.
527. Fujioka T, Furumi K, Fujii H, et al. Antiproliferative constituents from umbelliferae plants. V. A new furanocoumarin and falcarindiol furanocoumarin ethers from the root of Angelica japonica. *Chem Pharm Bull (Tokyo)* 1999;**47**:96–100.
528. Kim NY, Pae HO, Ko YS, et al. In vitro inducible nitric oxide synthesis inhibitory active constituents from Fraxinus rhynchophylla. *Planta Med* 1999;**65**:656–8.
529. Kang TH, Pae HO, Jeong SJ, et al. Scopoletin: an inducible nitric oxide synthesis inhibitory active constituent from Artemisia feddei. *Planta Med* 1999;**65**:400–3.
530. Farah MH, Samuelsson G. Pharmacologically active phenylpropanoids from Senra incana. *Planta Med* 1992;**58**:14–8.
531. Yun BS, Lee IK, Ryoo IJ, Yoo ID. Coumarins with monoamine oxidase inhibitory activity and antioxidative coumarino-lignans from Hibiscus syriacus. *J Nat Prod* 2001;**64**:1238–40.
532. Shaw CY, Chen CH, Hsu CC, Chen CC, Tsai YC. Antioxidant properties of scopoletin isolated from Sinomonium acutum. *Phytother Res* 2003;**17**:823–5.
533. Toda S. Inhibitory effects of phenylpropanoid metabolites on copper-induced protein oxidative modification of mice brain homogenate, in vitro. *Biol Trace Elem Res* 2002;**85**:183–8.

534. Shen Q, Peng Q, Shao J, et al. Synthesis and biological evaluation of functionalized coumarins as acetylcholinesterase inhibitors. *Eur J Med Chem* 2005;**40**:1307–15.
535. Hornick A, Lieb A, Vo NP, Rollinger JM, Stuppner H, Prast H. The coumarin scopoletin potentiates acetylcholine release from synaptosomes, amplifies hippocampal long-term potentiation and ameliorates anticholinergic- and age-impaired memory. *Neuroscience* 2011;**197**:280–92.
536. Cummings JL, Morstorf T, Zhong K. Alzheimer's disease drug-development pipeline: few candidates, frequent failures. *Alzheimers Res Ther* 2014;**6**:37.
537. Report of the JPND Action Group. *Experimental models for neurodegenerative diseases*; 2014.
538. Jucker M. The benefits and limitations of animal models for translational research in neurodegenerative diseases. *Nat Med* 2010;**16**:1210–4.
539. Gilman S, Koller M, Black R, et al. Clinical effects of Aβ immunization (AN1792) in patients with AD in an interrupted trial. *Neurology* 2005;**64**:1553–62.
540. Sriram S, Steiner I. Experimental allergic encephalomyelitis: a misleading model of multiple sclerosis. *Ann Neurol* 2005;**58**:939–45.
541. Hart BA, Gran B, Weissert R. EAE: imperfect but useful models of multiple sclerosis. *Trends Mol Med* 2011;**17**:119–25.
542. Lancaster MA, Renner M, Martin CA, et al. Cerebral organoids model human brain development and microcephaly. *Nature* 2013;**501**:373–9.
543. Xie YZ, Zhang RX. Neurodegenerative diseases in a dish: the promise of iPSC technology in disease modeling and therapeutic discovery. *Neurol Sci* 2015;**36**:21–7.
544. Guan ZZ, Yu WF, Nordberg A. Dual effects of nicotine on oxidative stress and neuroprotection in PC12 cells. *Neurochem Int* 2003;**43**:243–9.

Index

Note: Page numbers followed by *f* indicate figures and *np* indicate footnotes.

P

Q

R

S

T

Printed in the United States
By Bookmasters